THE CREATED COSMOS

THE CREATED COSMOS:

WHAT THE BIBLE REVEALS ABOUT ASTRONOMY

Danny R. Faulkner

with

Lee Anderson Jr.

Master
Books®
A Division of New Leaf Publishing Group
www.masterbooks.com

First printing: 2016

Copyright © 2016 by Danny R. Faulkner.
Appendix copyright © 2015 by Lee Anderson, Jr. Used by permission.

Master Books®, P.O. Box 726, Green Forest, AR 72638
Master Books® is a division of the New Leaf Publishing Group, Inc.

ISBN: 978-0-89051-973-8
Library of Congress Number: 2016910841

Front cover photograph by Jim Bonser. Used by permission.
Back cover photograph by Danny R. Faulkner.

Please consider requesting that a copy of this volume
be purchased by your local library system.

Printed in the United States of America

Please visit our website for other great titles:
www.masterbooks.com

For information regarding author interviews,
please contact the publicity department at (870) 438-5288

To the late George Mulfinger,
a professor of mine forty years ago.
We shared a passion for creation,
especially as it applied to astronomy.
He was a gentle man and a wonderful role model.

Contents

FOREWORD

One of the plagues in modern times with respect to the Bible-science stand-off is preachers who think they are scientists and Christian scientists who think they are Bible scholars. Not to be included are secular, humanistic scientists who pay no attention to the whole matter in any case, viewing it, as they do, from their Mount Olympus, as another spat of mere mortals who know nothing and have nothing worthwhile to say. Few indeed are the conservative, evangelical voices that can speak in the midst of the melee with authority founded on both credentialed scientific expertise and a degree of linguistic and exegetical skill sufficient to allow the sacred Scriptures also to speak and to do so with persuasive power. Such a work is this.

This writer, with like-minded others of late, has become greatly concerned about the inroads into contemporary evangelical scholarship of the assumption that the writers of the Hebrew texts followed lock-step the *Weltanschauung* and cultural norms of their respective historical environments, an adaptation by them to social and literary world-views that may be dubbed "patternism." This method presupposes a universal commonality of fundamental ideologies, unitary ways of reflecting on, interpreting, and recording the past, resulting, in Israel's case, in a certain degree of cultural adaptation and association, alleged literary borrowings and dependencies from secular sources by its scribes and scholars, and the like. Moreover, the pilfering of conceptual notions and the texts that relate them is always thought to be by the biblical writers, never the other way around. Thus, *Atra-ḫasīs* informs the Genesis creation story, *Enūma eliš* the biblical Flood narrative, and the Code of Hammurabi the Mosaic Torah.

Defenders of a "Bible uniqueness view" have found themselves foundering about in a morass of data, debate, and, sadly, defection from the time-honored views of the Fathers who preceded us and laid down for us an infrastructure of faith in the inerrant word that come what may was sufficient for every storm and conflict. One thinks of the Robert Dick Wilson, William Henry Green, Edward J. Young, O. T. Allis, R. Laird Harris, Gleason L. Archer, and others like them in yesteryear, men who stood firm against the tsunamis of liberal historical-criticism despite the risk of losing position, reputation, and academic renown in the eyes of those of lesser acumen and scholarship, counting it a small price to pay for truth and integrity.

In their wake have come evangelicals who, often well-trained and bright, could easily have taken up the mantle of the fallen warriors but have elected instead to belong to the academy where everyone knows and admires one another and where there is the ease and comfort of going with the flow. Why fight when you can join? The weapons of today are the new hermeneutics, certain brands of discourse analysis, aversion to literal readings of texts that speak of the unique, the unrepeatable, the miraculous, the supernatural, the non-documentable by modern tools of research; the fear of being caricatured as "Luddite," "medieval," "literalistic," "unscientific," "non-enlightened," or, worst of all, "fundamentalist." No wonder the ship of classical biblical scholarship is being abandoned. Who wants to be aboard a vessel so much off course that it is bound eventually to be grounded on the shoals of irrelevance and absurdity?

Danny Faulkner has chosen to remain on the ship as it were and to argue, on the basis of both astronomical and exegetical evidence, for a young earth and six-day creation, with an eye to all the varying rebuttals and responses he knows his case will surely engender. But this is a case constructed not by a novice or dilettante but by a credentialed scientist with a doctoral degree in astronomy. The amassing of astronomical data relevant to various theories as to the age of the earth, a geocentric galaxy or universe, the uniqueness of humankind in such a universe, causation, time and space, astronomical phenomena such as eclipses, "falling stars," the "long day" of Joshua, the "backward movement" of Hezekiah's sundial—all are treated fairly from all sides and accounted for astronomically, exegetically, and theologically. Rarely is the discipline of astronomy brought up in the so-called "Bible-science debate," most likely because of a lack of expertise in its complexities by both sides in the controversy. Close attention to the points raised here will help the serious reader understand the debate, to some degree at least, even if he or she cannot follow the technicalities naturally inherent in such an arcane field of study.

The Bible asserts that "God created the heavens and the earth" (Genesis 1:1) and he said moreover, "Let there be lights in the expanse of the heavens" and "let them be for signs and for seasons, and for days and for years" (Genesis 1:14). David understood the importance of this creative work, which he took literally to be the work of God alone: "The heavens declare the glory of God, and the sky above proclaims his handiwork" (Psalm 19:1). But this is not enough. The poet goes on to note that this kind of proclamation is insufficient because in astronomy "There is no speech, nor are there words; it has no voice" (verse 3, my translation). Thus there follows David's testimony about Scripture:

> The law of the Lord is perfect,
>   reviving the soul;
> the testimony of the Lord is sure,
>   making wise the simple.

Astronomy speaks a powerful word about God as Creator, if only one has eyes to see; the Bible speaks a powerful word about God as Re-creator and Redeemer, if only one has faith to believe.

EUGENE H. MERRILL
JUNE 9, 2016

ACKNOWLEDGMENTS

A tremendous amount of credit goes to Lee Anderson for getting this book finished. I have thought of writing this book for at least four years. However, it wasn't until I recently mentioned the book to Lee and he got excited about the prospect that this project took off. After discussion of the topics that the book ought to cover, Lee organized the material and presented it to Dr. Andrew Snelling, the Director of Research at Answers in Genesis. Andrew enthusiastically approved the book as a Research Department project, and we soon began work. As I began to write, Lee and I frequently met to further organize the book's contents. When I would finish drafts of each chapter, I would send them to Lee for his helpful editorial suggestions. For Chapter 1, our roles were reversed—Lee wrote the first draft and I offered suggestions. However, I must confess that I had very few suggestions, far fewer than he had for my writing. In conjunction with the first chapter, Lee also wrote the Appendix. Even more importantly, I found Lee's assistance extremely helpful in correctly understanding elements of biblical Hebrew and in getting recommendations on theological issues. Not only did Lee's organizational skills greatly speed up the writing process, the overall quality of the book is far better than I could have managed on my own.

There are other people that I ought to mention. As previously stated, Andrew Snelling, approved this as an Answers in Genesis Research Department project. This also greatly sped up the writing process, so I thank both Andrew and Answers in Genesis for their support. My colleague and good friend, Dr. Robert Hill, was helpful as a sounding board and in reviewing the entire book. Not only is Bob a fellow astronomer, he brings a broad breadth of knowledge in such a project. Steve Golden did the general proofreading of the entire book. Having worked with Steve for several years, I have learned that he is a master of writing, so I am grateful for his help in this.

Though he may not have been the original source of this sentiment, Sir Isaac Newton once wrote, "If I have seen further, it is by standing on the shoulders of giants." I truly understand this. I have had the pleasure of knowing personally the co-founders of the modern creation movement, Dr. John C. Whitcomb and the late Dr. Henry M. Morris. These two giants have influenced either directly or indirectly virtually all recent creationists in the world today. But there have been other giants that have helped me see further, such as the late George Mulfinger, a professor, example, and mentor of mine 40 years ago. I am sure that he would have been happy with this book.

One other giant in the realm of theological study is Dr. Eugene Merrill, who graciously agreed to write the foreword for this work. I am extremely grateful for his kindness and support.

Finally, I wish to thank publicly my loving wife, Lynette. Over the years, she has encouraged me in so many ways. She has always been supportive of my calling (or shall I say, *our* calling) as an astronomer for God's glory.

DANNY R. FAULKNER
APRIL 22, 2016

Astronomy and the Study of the Bible

In creation science, we discuss various sciences within the context of the biblical doctrine of creation. The biblical doctrine of creation in turn is gleaned from scriptural passages that deal with the creation event. Unfortunately, it sometimes seems that we start with the study of science, working to build our models within the framework of the doctrine of creation, and only thereafter asking what the Bible has to say about it. This presents the danger that our creation model as a whole may be a little more inclined toward science and a little less inclined toward Scripture than it ought to be.

In endeavoring to correct this tendency, we do well to consider the disciplines of biblical and systematic theology, respectively. Biblical theology is the study of the doctrinal content of the Bible, treating the Bible as the only source, examining the content of each individual book of the Bible (or group of books by a single author) within its own historical context. Systematic theology is the attempt to organize biblical truth categorically, crossing over the boundaries of authorship and historical context, and then, to use the resultant understanding of biblical truth as the standard for either affirming or rejecting truth claims arising out of other disciplines, including science.

Given this definition, it is appropriate to view the work of creation scientists today as taking place within the broad sphere of systematic theology, in that they attempt to understand their respective fields within the parameters defined by Scripture. This approach is a good one, as long as we keep our science rooted in what the Bible *actually* says. If there is a disconnect between the careful interpretation of the biblical data (in the study of biblical theology) and the interaction with the truth claims coming out of the study of the sciences (in the realm of systematic theology), we lose our basis for confidence in the accuracy of our scientific models, as Scripture no longer controls the parameters within which we are seeking to construct those models. Therefore, it behooves us from time to time to have a discussion of a science, such as astronomy, that *explicitly* starts with the Bible to see what the text says about it. While there are a number of treatments of astronomy as it relates to creation, there are few, if any, current resources that simply and specifically begin by asking what the Bible reveals about astronomy. There is no shortage of material, for the Bible contains an amazing number of mentions of astronomical concepts.

While there appears to be a lack of material today that specifically addresses a biblical outlook on astronomy, this has not always been the case, for a century ago there were several such books. One example is *Astronomy in the Old Testament* published in 1905 by the famous Italian astronomer Giovanni Schiaparelli. This was a translation of the Italian edition that appeared in 1904. Schiaparelli is best remembered for his observations of the planet Mars. Like other astronomers of his day, Schiaparelli thought that there was water on Mars, so he interpreted Martian features in this manner. In mapping the Martian surface, Schiaparelli included seas (darker areas) and continents (lighter areas).[1] During the great opposition of Mars in 1877, Schiaparelli saw linear features on the Martian surface that he called channels, suggesting possible natural connections between bodies of water. The Italian word for channels is *canali*, which unfortunately was mistranslated into English as "canals." This inspired a popular fascination with Mars especially in the United States, and soon caught up the American astronomer Percival Lowell, who spent the rest of his life promoting his belief that canals on Mars proved that life existed there. It turned out that the channels on Mars were optical illusions. However, Schiaparelli did far more significant work than his discovery of what he thought were channels on Mars. For instance, he studied binary stars, and he was the first to show that meteor showers are caused by debris of comets, a view that eventually won wide acceptance. Schiaparelli also was an expert on the history of ancient astronomy. His work in this area dovetailed very nicely with his interest in biblical astronomy.

---

[1] Today we know that bodies of water do not exist on Mars, but that there may be a large amount of ice, frozen water, in the soil and rocks of Mars.

Another example of a work that specifically addresses a biblical outlook on astronomy is *Astronomy of the Bible: An Elementary Commentary on the Astronomical References in the Holy Scripture*, published in 1908 by E. Walter Maunder, an English astronomer best known for his study of sunspots. Maunder, along with his wife, Annie, was the first to show that the latitudes of sunspots varied throughout the sunspot cycle. He also showed from historical records that there was a 75-year period of very low sunspot activity in the late seventeenth and early eighteenth centuries, a period of time that we now call the Maunder Minimum. The Maunder Minimum and other periods of low sunspot activity correlate well with times of lower temperatures on earth, so it has become an important topic in the debate over global warming. In 1890, Maunder founded the British Astronomical Association, an organization primarily for amateur astronomers. It remains one of the premier amateur astronomy organizations in the world today. Maunder was an early critic of Lowell's theory about life on Mars, being one of the first to argue the Martian canals were optical illusions. In 1913, Maunder tackled the question of extraterrestrial life in his book *Are the Planets Inhabited?* He could have taken a biblical approach (see chapter 12 of this book), but Maunder instead took a purely scientific approach.

The two books by Schiaparelli and Maunder stand out for the breadth of their coverage of what the Bible says about astronomy. While these two authors appear to have had a commitment to the cardinal doctrines of Christianity and they handled some issues very well, on other issues their books are a disappointment. For instance, they both believed that there was no physical miracle at the Battle of Gibeon. Rather, they argued that the account of Joshua's long day in Joshua 10 was poetic license.

In 1919, Lucas A. Reed published *Astronomy and the Bible*, though the discussion contained in this book is not nearly as deep as the treatments of Schiaparelli and Maunder. Reed was a Seventh-day Adventist, and some Adventist ideas come through in his book. For instance, Reed discussed the "open space in Orion" that Ellen G. White wrote about, which forms the basis for Orion being the dwelling place of God, a common belief among Seventh-day Adventists. Furthermore, Reed concluded from Jeremiah 33:20–21 and Psalm 89:34–37 that the sun and moon will continue on into eternity.[2]

---

[2] It is interesting that Henry M. Morris also taught that the stars would continue on into eternity, though he primarily used Daniel 12:2 as his support.

An earlier example of a work addressing astronomy in the Bible was the 1871 posthumous publication of *The Astronomy of the Bible* by Ormsby Macknight Mitchel. Mitchel led a very interesting life. A graduate of West Point (in the same class as Robert E. Lee), Mitchel reentered civilian life upon completion of his military duty. He moved to Cincinnati, where he soon passed the bar, became a professor at Cincinnati College (now the University of Cincinnati), and was named the chief of engineering at the Little Miami Railroad. As if none of these three professions were enough, Mitchel soon founded the Cincinnati Observatory, the oldest professional observatory in the United States. Mitchel devoted an enormous amount of his personal time to this project. To raise the considerable funds for construction, Mitchel sold memberships, an unheard-of approach at the time. In fact, he traveled to Germany to purchase a suitable telescope, and the manufacturers were stunned by this method, because they were accustomed to working with royal patronage. To spark interest in the project, Mitchel gave a series of popular lectures. He must have been quite an orator, for only 16 people attended his first lecture, but soon the crowd swelled to 2,000. The text of these lectures eventually was collected in a book, *The Orbs of Heaven*, published in 1851. In these lectures, Mitchel made biblical references. His posthumous book, *The Astronomy of the Bible*, also a collection of popular lectures, went even further. In 1859, Mitchel departed Cincinnati for Albany, New York, to assume the position of astronomer at Dudley Observatory. With the beginning of the War Between the States, Mitchel resigned that position to take a commission as a brigadier general in the Union Army. In 1862, troops under his command made a raid that became known as the Great Locomotive Chase, which eventually became the subject of two movies. Later that year, now Major General Mitchel contracted yellow fever and died while leading Union troops in South Carolina.

Mitchel's two books are different from Schiaparelli's and Maunder's. While their books are expositions of biblical passages related to astronomy, Mitchel discussed the wonders of astronomy with a biblical emphasis. However, his commitment to biblical authority is clear enough. For instance, he began his first book (his first lecture) by asking what the first man (Adam) saw as night fell upon him only 6,000 years ago. Elsewhere in the book, he again stated his belief that mankind was created just 6,000 years ago. In his second book, Mitchel clearly criticized Laplace's nebular hypothesis of the formation of the solar system. The basis of the critique was that the nebular hypothesis was a naturalistic explanation. Still, it appears that Mitchel believed in the day-age theory. Hence, while he thought that humanity was scarcely 6,000 years old, Mitchell thought that the world was much older.

Besides these works, there have been some books published in recent years that address various astronomical issues from a biblical perspective. One of the best in this sense has been *The Stars Speak: Astronomy in the Bible* by Stewart Custer, published in 1977. Much of the text of this book came from planetarium shows that Custer wrote for the Bob Jones University planetarium. Don DeYoung published *Astronomy and the Bible: Questions and Answers* in 1989 (revised in 2010). As the title suggests, this book is arranged in the form of questions (100 questions in total), followed by short answers (usually a page or less). In 1996, Werner Gitt published *Stars and Their Purpose: Signposts in Space*. Many of the chapters in this book have titles in the form of questions. In this respect, Gitt's book is similar to DeYoung's, albeit there are fewer questions and they are answered in more detail. Much of the information in Gitt's book is presented as a discussion of the wonders of creation, in that he used many facts demonstrating vast sizes and other large numbers found in astronomy. This is similar to the approach that Mitchel took. Stuart Burgess published *He Made the Stars Also: What the Bible Says about the Stars* in 2001. Much of Burgess' book is in response to issues, such as the big bang and extraterrestrial life, which seem to contradict a biblical worldview. Finally, Jason Lisle published *Taking Back Astronomy* in 2006. However, these books are arranged more around topics that are then addressed from Scripture, rather than first asking what the Bible has to say about astronomy, which was the approach of Schiaparelli and Maunder.

Also worthy of mention is Paul Steidl's 1979 publication *The Earth, the Stars, and the Bible*. Though a bit dated, this book remains the most comprehensive discussion of modern astronomy within a recent creation viewpoint. As excellent as Steidl's book is, it suffers from the mild criticism introduced at the beginning of this introduction—it approaches astronomy from the perspective of biblical creation rather than the approach of asking what the Bible has to say about astronomy. Apparently, while the latter approach once was popular, it no longer is. A book taking this approach has not been attempted in about a century, so the time may be right for a new book on the subject.

Of course, a book of this type probably ought not to be exclusively about biblical astronomy, because there are some relevant questions that may be omitted if one were to discuss *only* biblical astronomy. Hence, this book is divided into four parts. The first part is entitled "Astronomical Concepts in the Bible: What Scripture Plainly Teaches about Astronomy." It is the intent of the chapters in Part 1 to tackle a biblical outlook on astronomy as defined above. Part 2 is entitled "Astronomical Anomalies in the Bible: What Scripture Says about Unusual Astronomical Events." The purpose of the chapters in Part 2 is to investigate in more detail certain questions about astronomical issues that Scripture discusses. These include things such as the question of what the Christmas star might have been, or what happened with the miracles of Joshua's long day or Hezekiah's sundial. Part 3 is entitled "Astronomical Questions and the Bible: How Scripture Confronts Recent Questions about Astronomy." The three chapters of Part 3 discuss some issues concerning the Bible and astronomy, such as the light-travel-time problem, which have been raised relatively recently. While the Bible does not directly address these questions, these are questions about the Bible that people naturally have, so it would be remiss to omit discussion of these topics in a book of this sort. Finally, Part 4 is entitled "Astronomy and Distortions of the Bible: Misconceptions of What Scripture Teaches about Astronomy." The chapters in Part 4 respond to claims that various people make about what the Bible supposedly teaches, but actually doesn't teach. Some of these ideas come from skeptics, such as the idea that the Bible allegedly teaches that the earth is flat. Others ideas come from very devout Christians, such as the proposed "gospel in the stars." I hope that the reader can see the logic of this ordering, going from what the Bible clearly teaches, to things that we can glean from Scripture, to answering questions with biblical principles even though Scripture does not directly address those questions, to refuting claims about what the Bible teaches when it actually does not.

Finally, it is my sincere desire that those reading this book will be encouraged and edified by it. In the words of Johann Sebastian Bach, *Soli Deo gloria.*

PART 1

Astronomical Concepts in the Bible: What Scripture Plainly Teaches about Astronomy

The Bible contains a number of mentions of, or allusions to, astronomical bodies and concepts. The very first words of Scripture speak of the creation of the heavens and the earth. Right away, this sweeping statement introduces the field of astronomy, though the astronomical bodies were not made until Day Four of the Creation Week, several verses into the creation account of Genesis 1. In between, on Day Two, God made the firmament, or expanse. This immediately raises all sorts of questions, such as, *What is this firmament? What was the source of light for the first three days? And how can we see distant starlight?* However, this is getting ahead of ourselves.

This part of the book begins with a broad discussion of biblical theology and its relationship to astronomy in Chapter 1. Psalm 19:1 makes it abundantly clear that what we see in the heavens above demonstrates God's glory. As such, the study of astronomy is a God-honoring endeavor. I have always thought that I am truly blessed in my work as an astronomer, for it is a holy calling. Most people are awed by images of astronomical bodies. I think that there is a good reason for this: Astronomy, as Psalm 19:1 indicates, gives each of us a connection to our Creator. I don't understand people who don't find astronomy fascinating.

Chapters 2, 3, and 4 continue this study first with the Bible's teachings on heaven and earth, followed by the Bible's teachings about the sun, moon, and stars, and then a discussion of the purposes of the heavenly bodies. Chapter 5 concludes Part 1 with another clear message from Scripture related to astronomy—biblical warnings against astrology. Astrology is an ancient pagan religion, but unlike other ancient pagan religions, it survives well today.

Astrology is not a simple diversion, but instead represents a system that works to detract our attention from the true God.

In many respects, this first part is the most important portion of this book, because it lets the Bible speak for itself on astronomical matters. The remaining three parts will progressively move away from these clear teachings to the consideration of less clear issues.

CHAPTER 1

Biblical Theology and the Study of Astronomy

Lee Anderson Jr.

Astronomy, one of the oldest of the sciences, is the study of objects and matter outside of earth's atmosphere, and of their physical and chemical properties.

The Scriptures are limited in what they communicate about astronomy; however, what the Bible does tell us about astronomy is of unparalleled importance. From the Scriptures we know that the heavens and all they contain were created by God (e.g., Gen. 1:14–19) and that they serve ultimately to magnify His glory (Ps. 19:1). The tremendous vastness of the heavens, clearly implied by the Lord's rhetorical request that Abram "number the stars" (Gen. 15:5), ought to leave the observer overawed by the power and majesty of the Sovereign Lord who spoke the astronomical realm into existence. Fittingly, David surmised in Psalm 8:3–4,

> When I look at your heavens, the work of your fingers,
>   the moon and the stars, which you have set in place,
> what is man that you are mindful of him,
>   and the son of man that you care for him?

Given the splendor of God's heavenly creation, it is not surprising in the least that David was compelled to conclude this psalm with the exclamatory refrain in verse 9,

> O Lord, our Lord,
>   how majestic is your name in all the earth!

In view of the stated purpose that the heavens have in magnifying the glory of the living God, we are correct to conclude that the study of astronomy is a worthy and God-honoring endeavor. This conclusion is supported by prominent biblical examples. For instance, the wisest man to ever live, King Solomon, is mentioned favorably in 1 Kings 4:33–34 for devoting himself to scientific pursuits; and, as we saw above, David gave thoughtful consideration to the workings of the heavenly bodies. Thus, we see that the discipline of astronomy is a worthwhile field of study, and that the pursuit of knowledge in this area of science is a noble goal.

In approaching the study of astronomy within a Christian worldview, our thinking on the subject must be governed first and foremost by what the Scriptures teach. Because the Scriptures are "God-breathed" (2 Tim. 3:16) and wholly true in all they teach (Ps. 119:160; John 17:17), we are justified in looking to the biblical text as the appropriate starting point for the construction of our worldview.[1] As it concerns the study of astronomy, we look to the synthetic message of what the Scriptures proclaim about the cosmos to inform (and, sometimes, to constrain) the development of our scientific models. In this chapter, we will therefore seek to conduct a broad survey of what the Bible teaches about the heavens and the things therein, which will serve to frame the topics addressed in the chapters to follow. In particular, we will consider the Bible's grand purposes in discussing the wonders of God's created cosmos.

While we may grant, as stated before, that the Bible is limited in what it says about the study of the heavens—it is not an astronomy textbook—we can nevertheless be confident that the pages of Scripture will reveal a wealth of knowledge. With respect to all matters on which Scripture touches, we may hold tightly the promise of Psalm 119:130:

The unfolding of your words gives light;
  it imparts understanding to the simple.

That which perhaps figures the most prominently in the Bible's discussion of astronomy is the origin of the cosmos. Though succinct, the Scriptures are brilliantly explicit in ascribing the creation of the heavens to God. On Day Two of the Creation Week, detailed in Genesis 1:6–8, God formed the "expanse" and called that expanse "heaven." On Day Four, God spoke the heavenly bodies into existence, as it is detailed in Genesis 1:14–19:

---

[1] For a fuller discussion of the role of Scripture in the construction of a Christian worldview through theological method, and how it equips the Christian to engage with the truth claims of science (and other disciplines), see the Appendix: "Scripture as the Controlling Factor in Christian Worldview Development."

And God said, "Let there be lights in the expanse of the heavens to separate the day from the night. And let them be for signs and for seasons, and for days and years, and let them be lights in the expanse of the heavens to give light upon the earth." And it was so. And God made the two great lights—the greater light to rule the day and the lesser light to rule the night—and the stars. And God set them in the expanse of the heavens to give light on the earth, to rule over the day and over the night, and to separate the light from the darkness. And God saw that it was good. And there was evening and there was morning, the fourth day.

The creation of the two heavenly bodies most important to life on earth—the sun and the moon, described in this passage as "the greater light" and "the lesser light," respectively—receive the bulk of the attention in these verses. The stars are mentioned collectively, and are relegated to what is essentially a footnote in the text. The Hebrew word for "star" (כּוֹכָב) refers to any point of light in the heavens (including planets, distant galaxies, etc.), so there was no need in Genesis 1 to expressly mention any other astronomical phenomena. However, lest there be any doubt that the events of the Creation Week accounted for the formation of the entire cosmos, Scripture notes, "in six days the LORD made heaven and earth, the sea, and all that is in them" (Exod. 20:11). So too, Hebrews 11:3 states, "By faith we understand that the universe (αἰῶνας) was created by the word of God, so that what is seen was not made out of things that are visible."[2]

Here in the creation of the heavens is showcased the incredible power of the Lord. As if creating the earth and all that is in it were not enough, the Lord crafted a great expanse filled with billions of galaxies, each containing billions of stars, along with (potentially) countless planets, moons, and other entities. Scripture draws attention to the vastness of heaven, thereby amplifying its attestation to the Lord's awesome power. We have already noted the implications of God's rhetorical directive that Abram count the stars. Likewise, Job 22:12 maintains,

Is not God high in the heavens?
See the highest stars, how lofty they are!

---

[2] For discussion on the translation of αἰῶνας, see F. F. Bruce, *The Epistle to the Hebrews*, The New International Commentary on the New Testament (Grand Rapids, MI: William B. Eerdmans Publishing Company, 1964), 4, 280. Bruce convincingly argues that here is meant "the whole universe of space and time."

Though it is unlikely that the biblical authors could have fully grasped the incredible size of the heavenly expanse, they surely knew that the skies were of seemingly endless depth, with many of the celestial objects so distant that they were beyond any hope of man reaching them. The Lord, to have made outer space so vast, must have at His disposal unimaginably great power! This fact is underscored by Scripture's reminder to its readers that God was unaided in making the cosmos. Everything we observe in the night sky, from the most intricate details of Saturn's rings, to the artistic grandeur of the Milky Way, was the result of God acting according to His perfect wisdom and sovereign design. Isaiah 44:24 records the Creator's words:

> Thus says the LORD, your Redeemer,
>   who formed you from the womb:
> "I am the LORD, who made all things,
>   who alone stretched out the heavens,
>   who spread out the earth by myself."

Truly, the human mind is downright incapable of fathoming the essentially limitless power that must have been required to fashion the heavens. Human beings, even working together in large teams, struggle to engineer things which pale by comparison to even a single star, much less a galaxy full of them. The fact that the Scriptures employ an anthropomorphism, describing the creation of the stars as being merely the work of God's *fingers* (Ps. 8:3) ought to make us pause in humble, reverent awe of the Lord's omnipotence that was brilliantly displayed in His mighty work of creation.[3] As Psalm 33:8–9 fittingly exhorts its readers,

> Let all the earth fear the LORD;
>   let all the inhabitants of the world stand in awe of him!
> For he spoke, and it came to be;
>   he commanded, and it stood firm.

As much as God's power is displayed in His creative acts, though, we must also acknowledge the Bible's attestation to His power shown in how He

---

[3] As a point of devotional reflection, it is significant that while the stars are described as merely being the work of God's fingers, the text emphatically states that God bares His mighty arm to accomplish salvation for His people (Isa. 52:10; cf. 53:1). This emphasizes the care and attention that the Lord devotes to His redemptive work, and ought to be something in which we take great comfort. The God who displays unfathomable might in creation exercises His power all the more in bringing salvation to those who trust in Him.

has subsequently interacted with the cosmos. At the time of creation, God established natural laws that govern physical processes, including, among other things, the movement of the heavenly bodies. The movements of the earth, sun, moon, planets, and stars relative to each other follow a set pattern controlled principally by the force of gravity. These movements are reliable and mathematically predictable, and, as far as we are concerned, so regular that any deviation from their normal pattern would be cause for great surprise—if not outright amazement—as to the cause. Yet on at least two occasions in recorded history, the Lord miraculously intervened in the movement of the heavenly bodies in order to accomplish His purposes. In Joshua 10, the Lord apparently halted the rotation of the earth so that the Israelite army would continue to have daylight and be able to complete the rout of their enemies, the wicked Amorites. As such, the writer of Joshua concludes succinctly, "There has been no day like it before or since...the LORD fought for Israel" (Josh. 10:14).

Similarly, in 2 Kings 20 (cf. 2 Chron. 32; Isa. 38) the Lord caused a reversal of the earth's rotation such that the shadow cast on "the steps of Ahaz" progressed opposite to its normal direction. This miracle was given as a sign of the Lord's promise to heal Hezekiah from a deadly disease. In both cases, we observe that God altered the motion of at least one of the heavenly bodies, effectively suspending (by supernatural means) the physical laws at play. Despite efforts to account for these events in terms of purely natural phenomena, no compelling explanations have been set forth. Yet this is truly unnecessary. After all, we must realize that, for the God who created everything in the heavens, the miraculous suspension of the physical laws so as to bring about His purposes in these rare instances *is really nothing at all*.[4] (Both of these astronomical anomalies are discussed in more detail in Chapter 6 of this book.)

---

[4] Critics of the inerrancy of Scripture have pointed to these passages (especially Josh. 10) in order to argue that the Bible supports a geocentric view of the universe—which is clearly contrary to observation. After all, if the sun and moon are told to "stand still," this implies that the earth is stationary and is being circled by the other bodies. And, as the argument goes, if the Scriptures get the facts wrong in this instance, how can they be trusted in other matters? Notwithstanding the efforts of zealous (but grossly misguided) Christians to "prove" geocentrism in the hopes of "defending" the Bible, the fact remains that what we encounter in these two texts is quite recognizably just the very normal use of phenomenological language—rather than a scientific description—to explain the events. We often use such language today (far more often than we use technical scientific description) to express the same sorts of things. For example, in observing a beautiful sunset, no one exclaims, "Wow, wasn't that a lovely rotation of the earth!" The manner of expression must fit the context of the remark, and, as such, it is perfectly natural (and correct) for the Bible to speak of the halting of the earth's rotation as the sun and moon standing still.

Returning now to the creation narrative, we find embedded in the Genesis account of the Lord's creation of the cosmos two very closely related polemics. While these polemics may seem hardly evident to us reading English translations of the biblical text in the present day, they would have been conspicuously apparent to the original readers of the Hebrew Scriptures when Moses wrote Genesis (15th century BC). These polemics evidence themselves when we compare the biblical record of creation to ancient Near Eastern creation myths. While it has become commonplace in contemporary biblical scholarship (particularly of a more liberal vein) to stress the *similarities* between the Bible and the ancient Near Eastern source material—and so conclude that they are not essentially different in their focus, message, or origin—the original readers of the biblical text would have in fact been struck by the profound *differences* which exist in the Scriptures. So this is what we will devote our attention to in the following paragraphs.

The first polemic that we observe in the creation account runs throughout the whole narrative, from Genesis 1:1 to 2:3. Notably, this polemic, though it would have been patently obvious to the original readers, to us seems subtle, because it concerns principally *what is not* in the text, rather than what is there. The ancient Near Eastern creation myths, such as the notable *Enûma Eliš*, are really more about the origin of the gods than the origin of the universe. In this sense, they are *theogonies*,[5] not *cosmogonies*. The ancient Near Eastern myths do not present the gods as existing eternally, but they are born out of eternal matter or are fashioned by other gods. These gods are often not very god-like, but are typically very petty and human in their character. While powerful, they are usually mere deified natural forces. Moreover, in these myths, the universe is typically made only after a period of battle between opposing deities, through which the victor essentially wins the right to create. But, even then, the victorious deity does not "create" the world in the sense of crafting it from nothing. Rather, he fashions the world from preexistent matter—in some cases, from the carcass of a vanquished rival deity.

Genesis stands in stark contrast to these crude and disgusting myths. In Genesis, the existence of the Lord is neither explained nor defended, but is rather assumed. This strongly implies His self-existence and eternality. The Lord is totally separate from His creation and is not associated with any kind of natural force. In His creative actions, the Lord speaks the creation into

---

[5] Theogony is the origin or genealogy of a group or system of gods.

existence in the span of six days; He relies upon no matter that existed before this time. Furthermore, the Lord never encounters resistance in His creative work; He exercises complete sovereignty in His work, unhindered by any opposing beings or forces.[6] Thus, the whole creation account, of which the Day Two and Day Four records of the creation of the cosmos are integral parts, shows the Lord in perfect dignity, strikingly different from the ancient Near Eastern deities.

The second polemic, which more directly concerns the Day Four account, again has to do mainly with what the text does not include rather than what it does include. As noted above, Genesis 1:14–19 describes the creation of the sun, moon, and stars. Notably, neither the sun nor moon are mentioned by name, but are simply referred to (collectively) as "the two great lights" and (individually) as "the greater light" and "the lesser light," respectively. This lack of naming is to show that the sun and moon are not gods, nor are they even sentient beings. Their stated purpose, "to give light on the earth" demonstrates implicitly that they have been placed in the heavens to serve man, which inherently precludes any notion that they are to be worshiped. (Notably, the Scriptures expressly forbid this elsewhere; e.g., Deut. 4:19.) Furthermore, since the Lord is shown to be the one who created the sun, moon, and stars, the only reasonable conclusion is that He is greater than they, so *even if* one insists that they are deities (which the text does not affirm) they are nevertheless subservient to the Lord. The Scriptures show that He is supreme over them, and they do not compare to Him.

The polemic against the sun and moon is picked up later in the biblical text. For instance, in Psalm 121:5–6, the psalmist utters these comforting words:

> The LORD is your keeper;
>   the LORD is your shade on your right hand.
> The sun shall not strike you by day,
>   nor the moon by night.

---

[6] For further discussion of the polemical features of the creation record, especially in how the Lord and His creative work are juxtaposed against the ancient Near Eastern deities and their efforts, see Steven W. Boyd, "The Genre of Genesis 1:1–2:3: What Means This Text?" in *Coming to Grips with Genesis: Biblical Authority and the Age of the Earth*, edited by Terry Mortenson and Thane H. Ury (Green Forest, AR: Master Books, 2008), 188–89. Particularly pivotal in this discussion are considerations of differences with respect to the reason for the creation of humanity and the place that is then given to humanity in the created order.

Even today, we might relate to the desire for shade from the sun; but the moon? The original readers of the psalm were hardly concerned with "moonburn" (or sunburn, for that matter). This text advances an attack on the idea, common some 3,000 years ago, that the sun and moon were deities—deities who might harm the children of Israel. No, says the God of Israel; no harm will come on account of the fabled sun and moon gods! The sun and moon are created entities; they are the Lord's handiwork (verse 2). Additionally, the Lord God is ever vigilant (verses 3–4), and will Himself ensure the protection of His people. It is with this assurance that the psalmist can rightly say that his help "comes from the LORD" (verse 2).

The point to take away from this is that the Bible, through its polemics, teaches something about astronomy that virtually no other ancient writings or cultures recognized: the heavenly bodies are not deities, nor are they to be worshiped as such. They are physical entities, created by the one true God, who rules over them in majestic, unopposed sovereignty.[7]

Having looked briefly at Scripture's teaching about the creation of the cosmos, as well as at the role that the Day Four creation narrative plays theologically and polemically, we now turn our attention to the Bible's discussion of astronomy as it concerns the structure of the cosmos. It is in this area especially to which the Bible speaks but little, but it does still touch on the matter. And, while the Scriptures do not address this topic the way we might expect of a science textbook, we still see that what Scripture says about astronomy is entirely concordant with that which we have learned from

---

[7] There are, of course, more direct polemics in the text directed against sun and moon deities that communicate their message through what they do say, rather than what they omit. For example, in Exodus 10, the Lord sends a plague of darkness on Egypt, making the land pitch black for three days and nights. It was a darkness so severe that, as the text says, it could be felt (Exod. 10:21). The darkening of the skies—especially during the daytime—was a not-so-subtle attack on the Egyptian sun god *Ra*, and the plague would have been understood as the Lord doing battle for His people, whom the Egyptians had enslaved and oppressed. Interestingly, this plague is more integral to the narrative than many people realize, as can be shown by its relationship to the plague to follow. In ancient Egyptian mythic lore, *Ra* is the father figure to Pharaoh. The plague of darkness showed *Ra* to be powerless before the Lord, incapable of defending Pharaoh or the Egyptian people. In the tenth plague, recounted in Exodus 12, the firstborn children throughout the land of Egypt die, including (and explicitly mentioned, lest it be overlooked) "*the firstborn of Pharaoh who sat on his throne*" (Exod. 12:29). By means of these two plagues, the Lord had cut Pharaoh off from both his fabled heritage and his actual progeny. He was left alone and vulnerable. It was almost as if the Lord was saying to Pharaoh, "I'm coming for you next, and there is nothing anyone can do to stop Me."

observational science. In much of what it says about astronomy, the Bible has been shown to be ahead of its time compared to other ancient sources.

As we have mentioned already, the Bible indicates that the universe is extremely vast and that it contains more stars than can be identified and numbered (Gen. 15:5; 22:17; Deut. 1:10; 10:22). This claim of Scripture stands in contrast to the common belief of a bygone era in which it was erroneously assumed that Ptolemy's catalogue of 1,022 stars in his 2nd century BC work, *The Almagest*, was exhaustive. (Interestingly, Ptolemy never claimed his work was exhaustive.) Of course, observation of the night sky with even a basic telescope or a good set of binoculars is enough to corroborate the testimony of Scripture, and the large telescopes used by professional astronomers have shown without a doubt that the number of stars in the universe may indeed be, as the Bible says, likened to the number of sand grains on the shore.

Aside from this point, in Judges 5:20, Scripture also acknowledges the "courses" of the heavenly objects, that is, their pattern of motion through the sky. While we must recognize the figurative language used in reference to the stars in this verse—which is appropriate to the poetic genre of the passage— there is no satisfactory reason *not* to understand the mention of the "courses" of the stars as a simple but accurate description of the mechanics of their motion. While other ancient cultures understood the predictable nature of the heavens and were able to track the motion of the heavenly bodies, the Bible goes further to give the reason for *why* such tracking and predication was possible. In Jeremiah 33:25, the Lord states that He had made a covenant with day and night, and that He had established "the fixed order (חֻקּוֹת; *literally*, statutes) of heaven and earth." This, it seems, is a subtle reference to the physical laws that govern planetary motion—laws that we did not really begin to understand scientifically until millennia later.

One final noteworthy point about the structure of the cosmos concerns Job 26:7, in which we are told that God "hangs the earth on nothing." While some ancient cultures conceived of the earth as being supported by some giant object (e.g., a tortoise), Job rightly presents the earth as being suspended in space. As a word of caution, we must be aware of the fact that much of the book of Job is poetry and avoid unnaturally forcing its poetic statements into an overly literalistic mold, taking them as scientifically exacting assertions about cosmology. Indeed, if we were to do this, how would we read other poetic—

and clearly figurative—expressions in Job which speak of the earth as resting on "pillars" (Job 9:6), or on a foundation with "bases" and a "cornerstone" (Job 38:6)? Job 26:7 should not be made to say more than the biblical author intended. At the very least, however, it may be affirmed that Job 26:7 denies a mythical cosmology, and in doing so it *allows* for observational science to inform one's approach to cosmology. R. Laird Harris' outlook on this matter is appropriately balanced and perceptive. He writes,

> Job plainly says that God "hangs the earth upon nothing" (Job 26:7), which gives not detail, but is in accord with the facts and faithfully avoids the bizarre and mythological pictures that were sometimes used in antiquity. The Bible was written in an early age. But it was written by God who created the world and who knows the end from the beginning. No one needs to reject the Bible because it is alleged to contain an outmoded view of the world.[8]

In summary, therefore, we see that while the Bible speaks little about the layout of the heavens, what it does say is truly profound and wholly worthy of our attention.

This now leads us to consider what the Scriptures say about the *function* of the cosmos and the objects therein. As we saw earlier in the Day Four creation account, God created the sun, moon, and stars "to give light on the earth" (Gen. 1:17). In the case of the sun, its brilliant light is accompanied by an appreciable amount of warmth (cf. Ps. 19:6), which is necessary for life to exist on earth. The sun especially, therefore, is a display of God's merciful providence to all the people of the world, whether they acknowledge Him or not (Matt. 5:45).

The Lord also states in Genesis 1:14 that the heavenly bodies are to serve as markers "for signs and for seasons, and for days and years." Chapter 4 will discuss in much more detail the function of the sun, moon, and stars for

---

[8] R. Laird Harris, "The Bible and Questions of Cosmology," *Presbyterion* 7, no. 1–2 (1981): 201. Another matter related to astronomy on which the language of the biblical text is commonly overextended is the expansion of the universe. It is often claimed that Isaiah 40:22 speaks of the present expansion of space, a scientific phenomenon that was not understood until the 20th century. However, every other verse in the book of Isaiah which speaks about the "stretching out" of the heavens (42:5; 44:24; 45:12; 48:13; 51:13; cf. Jer. 10:12; 51:15; Zech. 12:1) does so in reference to God's initial work of creating the universe. While it is true that in some sense God stretched out the heavens, this action on His part seems to have been relegated to the past. Thus, the Bible does not appear to speak directly to the presently observed expansion of space, though the text does not deny it either.

keeping track of time. However, it is worth noting that, even with the invention of highly accurate timekeeping devices, the determination of the day and year are still tied to astronomical indicators: the earth's daily rotation relative to the sun and the earth's annual revolution around the sun, respectively. Throughout much of history, most cultures used the cycles of the moon as the basis for determining the length of the month. Notably, the Hebrew word used in the Old Testament to indicate a month is חֹדֶשׁ, which literally means "new moon." Indeed, time measurement is a continuing function of the heavenly bodies.

Instances of the heavenly bodies serving as "signs" are comparatively rare in Scripture. The unusual events occurring in the days of Joshua and Hezekiah mentioned before constitute signs. However, the vast majority of astronomical signs mentioned in Scripture concern events which are yet to occur in the end times. For example, the prophet Isaiah links the approaching Day of the LORD with a mass darkening of the sun, moon, and stars (Isa. 13:9–10; cf. 24:21–23). Joel 2:30–31 echoes this theme, wherein the Lord describes coming judgment:

> "And I will show wonders in the heavens and on the earth, blood and fire and columns of smoke. The sun shall be turned to darkness, and the moon to blood, before the great and awesome day of the LORD comes."

Similarly, Jesus Christ in the New Testament announced that miraculous signs in the heavens would accompany His Second Advent. Matthew 24:29–30 (cf. Mark 13:24–26; Luke 21:25–27) describes in brief the wonder of these future events:

> "Immediately after the tribulation of those days the sun will be darkened, and the moon will not give its light, and the stars will fall from heaven, and the powers of the heavens will be shaken. Then will appear in heaven the sign of the Son of Man, and then all the tribes of the earth will mourn, and they will see the Son of Man coming on the clouds of heaven with power and great glory."

The book of Revelation, which arguably contains the most detailed biblical account of the events of the eschaton, mentions occurrences very similar to those foretold by Jesus Christ. It describes the future darkening of the sun, the moon becoming like blood in appearance, and the stars falling from their places (Rev. 6:12–14; cf. 8:12). After fulfilling their purposes in the Lord's spectacular eschatological judgments, ultimately the material heavens will pass away (Ps. 102:25; Heb. 1:10–12), being, as Isaiah 34:4 says, rolled up like a scroll. John, in his vision of the New Heaven and New Earth, states that the holy city, the New Jerusalem, has no need for the light of the sun or of the moon (implying that they do not exist), "for the glory of God gives it light, and its lamp is the Lamb [Jesus Christ]" (Rev. 21:23). These and other relevant eschatological passages will be discussed later in Chapter 9 of this book.[9]

The main point that may be seen in the grand sweep of what the Bible says about the heavens and the heavenly bodies is this: From the beginning to the end, from the original creation to the new creation, the cosmos and everything in them function to serve the purposes of the Lord and to evidence His greatness and power. The heavens are not merely incidental to the Lord's plan and purposes, but are integral to them, having been designed with tremendous wisdom and care. Thus, the psalmist does not call out in vain when he exclaims in Psalm 148:3–5,

---

[9] One other important function of the heavenly bodies which is not stated in the biblical text, but which still connects very closely with our study of the Bible, concerns the development of a biblical chronology. The Old Testament contains a fairly extensive *relative chronology*, in that it tells us when events happened relative to each other, but it does not provide an *absolute chronology*, one in which the events are connected with actual dates. However, placing the relative chronology offered by the Scriptures within an absolute chronology is enabled by observing correlations between astronomical events, which are mathematically calculable, and the Assyrian eponym (or *limmu*) lists, which provide centuries of unbroken sequences of years, with notes on the events of those years. Descriptions of an eclipse (known to have occurred in 763 BC) allow for the precise dating of multiple events in ancient Near Eastern history, including the Battle of Qarqar. From both biblical and extrabiblical data, we know that this battle must have occurred late in the reign of King Ahab, just before his death (cf. 1 Kings 22), for this was the only time during his reign in which Israel allied with its longtime enemy Syria (due to the encroachment of Assyria). This allows us to anchor the Bible's history within an absolute timeline, with dates accurate to within a few months. By employing a proper understanding of Israel and Judah's regnal dating practices, we can work our way back chronologically from the Battle of Qarqar to the reigns of Israel's first monarchs. Of greatest significance is Solomon, whose reign spanned 40 years, and commenced in 970 BC. From the plain statement of 1 Kings 6:1, we know that the fourth year of Solomon's reign was in the 480th year after the Exodus, which allows us to place the Exodus in 1446 BC. Looking to passages such as Exodus 12:40 (cf. Gen. 15:13; Acts 7:6), which indicate that the length of the Egyptian sojourn was 430 years, we are likewise able to accurately date the entrance of Israel into Egypt. And from that point, by relying on the chronological information in Genesis, we can also reconstruct the timeline of the Patriarchs. For more detailed discussion on this point, see Andrew E. Steinmann, *From Abraham to Paul: A Biblical Chronology* (Saint Louis, MO: Concordia Publishing House, 2011); and Eugene H. Merrill, *Kingdom of Priests: A History of Old Testament Israel*, 2nd ed. (Grand Rapids, MI: Baker Academic, 2008).

Praise him, sun and moon,
  praise him, all you shining stars!
Praise him, you highest heavens,
  and you waters above the heavens!
Let them praise the name of the Lord!
  For he commanded and they were created.

Though without voice, the heavens and the heavenly bodies do "praise" God by functioning harmoniously within their intended purpose. In fact, their very existence—along with the rest of creation—testifies to the existence and boundless might of an intelligent Creator. The intricacy, order, and grandeur of the cosmos are a clear witness to their Designer. In following this line of reasoning, the Scriptures offer a profound apologetic in Romans 1:18–20:

> For the wrath of God is revealed from heaven against all ungodliness and unrighteousness of men, who by their unrighteousness suppress the truth. For what can be known about God is plain to them, because God has shown it to them. For his invisible attributes, namely, his eternal power and divine nature, have been clearly perceived, ever since the creation of the world, in the things that have been made. So they are without excuse.

It is really inconceivable as to how the heavens—in all their vastness and splendor, and yet, at the same time, in their fine-tuned precision—could have come to exist without an intelligent Creator. All theories of cosmic evolution ultimately stumble over the question of a *first cause*. Moreover, assuming an evolutionary origin for the cosmos—which has *chaos* as its driving force and *chance* as its mastermind—it is virtually impossible to account for the remarkably precise order of our place in the cosmos. For example, our planet, earth, orbits the sun at just the right distance and has just the right axil tilt to sustain life. Our moon is of just the right size and mass, orbits earth at just the right distance, and exerts just the right amount of gravitational force on the planet to provide for healthy ocean tides. Our sun—when compared with the many other stars in our galaxy—has just the right mass and composition, and is so remarkably consistent, that it is beneficial rather than detrimental to the flourishing of life on our planet. If any of these factors, along with many others, were off even slightly, life on earth could not exist.

However, the heavens do more than testify to the *existence* of their Creator; they also convey something—albeit something limited—about His Person.

Psalm 19:1–4 says,

> The heavens declare the glory of God,
>> and the sky above proclaims his handiwork.
> Day to day pours out speech,
>> and night to night reveals knowledge.
> There is no speech, nor are there words,
>> whose voice is not heard.
> Their voice goes out through all the earth,
>> and their words to the end of the world.

This passage captures the fact that the witness of the heavens, though not employing human language, is still a means of communication capable of powerfully conveying its point. There is "knowledge" that is revealed, and the heavenly witness is universal in scope.

The heavens, the psalm says, "declare the glory of God," that is, in a nutshell, God's marvelous power, His unfathomable wisdom, and, ultimately, His deity (cf. Rom. 1:20)—which entails a worthiness of honor and worship.[10] This connection is more clearly traced out in Revelation 4:11, in which the Apostle John records words of praise from a scene of heavenly worship:

> "Worthy are you, our Lord and God,
>> to receive glory and honor and power,
> for you created all things,
>> and by your will they existed and were created."

Here the Lord receives praise as Creator of the universe. The one who made the sun, moon, and stars is not a faceless, nameless deity; but rather,

---

[10] This point is underscored by comments on the psalm by Willem A. VanGemeren, who says that, for the psalmist, "Creation reveals the Lord's royal majesty and sovereignty (cf. Ro 1:19–20). It evokes a response of recognition of God's existence, majesty, and wisdom—and therefore of praise (cf. Ro 10:18).... The glory and wisdom of God are evident in the vastness of space.... For the psalmist 'space' is not empty but a revelation of God's creation of the magnificent heavenly bodies, which are characterized by radiance and regularity. The verbs 'declare' and 'proclaim' are participle forms, expressive of the continuous revelation of the heavens, and could be translated 'keep on declaring...; keep on proclaiming.' The wars and disturbances on earth often camouflage God's glory, as they divert attention away from the created heavenly bodies, which show more clearly God's majesty by their regularity and orderliness. He alone is the Creator, because the magnificence of the heavenly bodies confirms that they are all 'the work of his hands.'" Willem A. VanGemeren, "Psalms," in volume 5 of *The Expositor's Bible Commentary*, rev. ed., edited by Tremper Longman III and David E. Garland (Grand Rapids, MI: Zondervan, 2008), 214–15.

as Psalm 19 goes on to make abundantly clear, He is the one true God, the LORD, who has issued forth His word (verse 7), who is worthy of fear (verse 9), who is righteous and true in His judgments (verse 9), and who offers the hope of redemption (verse 14). Of course, these things cannot be gleaned from the witness of the heavens alone. However, the spectacular testimony to the Creator that is manifested in the night sky ought to humble the observer in the face of the Creator's power and deity, and then spur him to seek out the Creator who reveals Himself further, and more intimately, in the pages of Scripture. The desired end is that the one who sees the Lord's glory revealed in the heavens can speak to Him as the psalmist in verse 14, with words full of adoration and praise:

> Let the words of my mouth and the meditation of my heart
> be acceptable in your sight,
> O LORD, my rock and my redeemer.

CHAPTER 2

Biblical Teachings about the Heavens and the Earth

The Heavens (Firmament)

The word *heaven* or *heavens* is found in many places in the Bible, and it would seem to have an astronomical connection in many uses. In 273 usages in the Greek New Testament, the word οὐρανός is translated consistently as "heaven," though it is occasionally rendered in various English translations as "air" or "sky." When it appears with the preposition ἐκ, some English Bibles translate it as "heavenly."

Of course, the Old Testament was largely written in Hebrew, so the word for *heaven* or *heavens* used in the Old Testament is different from the Greek. The first occurrence of the word heaven in the Bible is found in Genesis 1:1, which records that "In the beginning, God created the heavens (שָׁמַיִם, *šāmayim*) and the earth (אֶרֶץ, *'ereṣ*)." In Hebrew, *šāmayim* is a plural noun. Whether *šāmayim* is rendered as a singular noun ("heaven") or a plural ("heavens") generally is a matter of the translator's preference. While the Hebrew word *šāmayim* is translated as a plural in Genesis 1:1 in versions such as the English Standard Version (quoted here), the New American Standard Bible, and the New International Version, some versions, such as the King James Version, translate *šāmayim* as a singular. While *šāmayim* appears 421 times in 395 verses of the Old Testament, it is the subject of a sentence only rarely, as in Psalms 19:1 and 50:6 (cf. Judges 5:4). While *šāmayim* normally is translated as "heaven" or "heavens," it occasionally also is translated as "sky." On one occasion, when it appears in construct with the Hebrew word הבְּרִי (from the root הבר, "to divide"), it is translated as "astrologer" (Isa. 47:13).[1] The corresponding Aramaic word שְׁמַיִן occurs 38 times, and it is translated either "heaven" or "heavens" each time.

The Hebrew word *šāmayim* refers to things above us. As such, it can have three possible referents. For convenience, we can call these the three heavens, though this terminology does not appear in the Old Testament. The first heaven is the near distance above us. Today we would call this the atmosphere, though the atmosphere is a modern concept, one that ancient people, including the Hebrews, would not have recognized. Clouds, birds, and precipitation are phenomenon associated with this first heaven. For instance, Psalm 104:12 refers to the birds of heaven and Isaiah 55:10 speaks of rain and snow coming down from heaven. The second heaven is the astronomical realm, what we today would call space. The Old Testament describes stars as being in heaven, in Genesis 22:17, for example. The third heaven is the abode of God. Psalm 115:3 states that "our God is in the heavens." The only place in the Bible where this distinction and enumeration of the heavens is alluded to is in the New Testament, in 2 Corinthians 12:2–4, where Paul briefly described his experience in the "third heaven."

Since the distinction and enumeration of the heavens is not clearly taught in Scripture, one must exercise caution in making these distinctions in biblical texts. The distinction is merely a phenomenological one. It is clear that things in the first heaven are nearby, because we readily can see motion in them, such as the motion of birds and clouds. Furthermore, there is an obvious parallax effect—one's location directly determines what one observes. A bird, a cloud, or rain experienced locally will not necessarily be experienced by observers elsewhere. On the other hand, a change in location on the earth's surface will not dramatically alter what we see in the second heaven, unless that change in location is great. Today we clearly see the distinction as being due to objects either being in the earth's atmosphere or beyond it, in space. However, the ancient Hebrews would not have grasped this distinction in this sense, because our concept of the atmosphere and space beyond is modern. Therefore, the distinction between the first and second heavens sometimes is blurred in the Old Testament, and they are merged into one in some contexts.

The second occurrence of the Hebrew word *šāmayim* is in the Day Two account of creation (Gen. 1:6–8). On Day Two, God made the expanse (רָקִיעַ; *rāqîaʿ*). Genesis 1:8 further states that God called the expanse "heaven"

---

[1] Appearing only once in the Hebrew Bible, the *ketib* (written) form is הָרְבוּ.

(*šāmayim*). Thus, we are in the curious position of God creating the heavens (or heaven) twice, in Genesis 1:1 and again in Genesis 1:6–8. There are several ways to resolve this issue, and the path that we take will have direct implications in our biblical cosmology. Important in this discussion is the *rāqîaʿ*, the thing that God made on Day Two and then called *šāmayim*.

The Hebrew word *rāqîaʿ*, translated as "firmament" in the King James Version, appears 17 times in the Old Testament, with over half of those occurrences (nine times) in Genesis 1. Given its prominence in the narrative of the Creation Week, understanding the *rāqîaʿ* is of inestimable value in developing a biblical cosmology. Since the word so rarely occurs outside of the creation account, discerning its meaning can pose a challenge. Recent creationists have developed several different understandings of what the *rāqîaʿ* is. For instance, Henry Morris, a co-founder of the modern creation movement, popularized the idea that the *rāqîaʿ* is the earth's atmosphere. However, Morris did allow that the *rāqîaʿ* in some passages might refer to astronomical heaven, or more properly, space itself. However, other creationists went further in identifying the *rāqîaʿ* with space in all contexts. For instance, in his 1994 book *Starlight and Time*, D. Russell Humphreys argued this and presented arguments against identifying the *rāqîaʿ* as the atmosphere. Walt Brown, best known for his hydroplate model, has a very different view, in that he contends that the *rāqîaʿ* is the earth's surface. All of these positions cannot be correct.

Different views of the *rāqîaʿ* lead to different conclusions about the earth, its atmosphere, and the universe, though it is not always clear if the views lead to the conclusions or if the conclusions require the views. For instance, the view that the *rāqîaʿ* is the atmosphere frequently leads to the conclusion that the pre-Flood earth was surrounded by some sort of water canopy. Indeed, the *rāqîaʿ* being the atmosphere was an integral part of Morris' Flood model, in that the collapse of the water canopy was posited as one of the two sources of the Flood waters mentioned in Genesis 7:11 (the "windows of heaven"). Through the considerable influence of Morris, by the 1970s the canopy model was widely popular among recent creationists. However, since then recent creationists have largely abandoned the canopy model. If the water canopy model is no longer viable, should we not reevaluate Morris' interpretation of the *rāqîaʿ*?

Still other creation scientists interpret the *rāqîaʿ* in a manner that is closely associated with a particular creation model. The *rāqîaʿ* being interstellar space is necessary for Humphreys' white hole cosmology. John Hartnett has attempted to explain the abundant water in the solar system by his identification of the *rāqîaʿ* with the space of the solar system. Separating the waters above and below by the earth's surface is an integral part of Brown's hydroplate model, with the *rāqîaʿ* being the primordial surface of the earth. Again, the fact that these beliefs about the *rāqîaʿ* contradict one another means that they all cannot be correct. Given the significant role that the *rāqîaʿ* appears to play in developing a biblical cosmology, it is very important that we properly understand what *rāqîaʿ* means.

Various translations of *rāqîaʿ* have originated and compounded the problem. The Septuagint translators chose to render *rāqîaʿ* as στερέωμα (*stereoma*). In ancient Greek cosmology, the *stereoma* was the hard, transparent sphere on which stars were affixed. As the *stereoma* spun, it carried the sun, moon, and stars across the sky. Of course, today we recognize that it is the earth's rotation that causes this motion. In most ancient Greek cosmologies there were other nested circles or spheres concentric within the *stereoma* that carried the sun, moon, and five naked-eye planets, producing motions of those objects with respect to the background of stars. The Septuagint translators' identification of the *rāqîaʿ* as the stereoma probably was an attempt to conform Scripture to the dominant cosmology of the day. The Septuagint translation was done in Alexandria, a center of Greek thought, and so the Greek influence was strong. The same appeal to conform to current thinking exists today, for many authors interpret Genesis 1 in terms of the big bang model, the dominant cosmological theory of our day.

In the Vulgate, Jerome chose the Latin word *firmamentum* to translate *rāqîaʿ*. As one easily may surmise, we get the English word *firm* from the root for this word, and so Jerome's choice here again went with the idea from the Septuagint of the *rāqîaʿ* being a hard substance. Many English translators, such as those of the King James Version, went along with Jerome by simply transliterating the Latin word as "firmament." Thus, the idea that *rāqîaʿ* denotes something hard persists among some creationists today.[2]

Given the reality of the way the *rāqîaʿ* has been translated, medieval Christian and rabbinical scholars' opinions on the subject may be suspect. At the very least they were products of the times in which they lived. Furthermore, they, like earlier translators, may have felt compelled to conform to the

---

[2] For instance, supporters of the hydroplate model often make this point and go on the equate the hard *rāqîaʿ* with the earth's surface.

cosmology of their times. Truly ancient (pre-Greek influence) Hebrew sources outside of the Old Testament are exceptionally rare and do not make mention of the word *rāqîa'*. Furthermore, the sparse use of the word *rāqîa'* elsewhere in the Old Testament is of little help.

The many ways that the biblical concept of the *rāqîa'* has been poorly handled has led to problems. For instance, it has become fashionable now to claim that the Bible's cosmology is that of a solid dome above the earth. This was a common ancient and even medieval concept, but has long since been rejected. Hence, if this is what the Bible taught, then that certainly would undermine the reliability of Scripture. But is this what the Bible teaches? No. Ancient and medieval attempts to interpret the Bible in terms of the current cosmology have led to this. Additionally, there has been an attempt to read the Old Testament in terms of ancient Near Eastern literature. The reasoning is that since at least some ancient Near Eastern cultures developed a solid dome cosmology, the Hebrews must have picked this up and incorporated it into the Old Testament. However, this line of reasoning relies upon what much later writers say about the Old Testament, not what the Old Testament actually says. Unfortunately, some Christians who ought to know better apparently have accepted this. For instance, the New International Version originally rendered the *rāqîa'* as "expanse," but since 2011, the updated New International Version has translated *rāqîa'* as "vault," thus endorsing this modern criticism of the Bible.

The Meaning of Genesis 1:1

The founders of the modern creation movement, John Whitcomb and Henry Morris, taught that Genesis 1:1 indicated that the creation of the earth and the space of the universe were the first creative acts of Day One. Given the stature that these two gentlemen have among creationists, it is not surprising that this has been the dominant view of recent creationists throughout the past half century. However, outside of the creation literature, this view of Genesis 1:1 is decidedly less dominant. Many Hebrew and Old Testament scholars (including conservative ones who believe in recent six-day creation) believe that Genesis 1:1 contains an example of *introductory encapsulation*. Introductory encapsulation amounts to a very brief summary that is followed by an elaboration of details. Consider this example:

Bill had a rough morning. His alarm did not go off, so he overslept. He didn't have time for breakfast, so he got hungry later. His car wouldn't start, so he had to jump start his car from his wife's car. The frost that morning was particularly heavy, so it took extra time to remove. Because he left much later than normal, the traffic was worse than usual. He arrived to work late for an important meeting.

The first statement, "Bill had a rough morning," is an example of introductory encapsulation. The encapsulatory summary is followed by six sentences elaborating the details of why Bill had a rough morning.

There are numerous examples of introductory encapsulation and elaboration found in the book of Genesis. Consider Genesis 37:5–8:

Now Joseph had a dream, and when he told it to his brothers they hated him even more.

He said to them, "Hear this dream that I have dreamed:

Behold, we were binding sheaves in the field, and behold, my sheaf arose and stood upright. And behold, your sheaves gathered around it and bowed down to my sheaf."

His brothers said to him, "Are you indeed to reign over us? Or are you indeed to rule over us?" So they hated him even more for his dreams and for his words.

Notice that the introductory encapsulation of the first sentence (Gen. 37:5) is a complete thought, though it lacks details. In verse 6, Joseph invites his brothers to hear his dream, and in verse 7 he gives the details of the dream. Verse 8 records Joseph's brothers' response and explains why their hatred of Joseph grew. Without this additional information, we would not know the reason for the brothers' increased hatred for Joseph.[3]

In similar manner, Genesis 1:1 functions as an example of introductory encapsulation, followed by elaboration given in Genesis 1:2–2:3. There are at least three reasons for understanding the text this way. First, the phrase "heaven and earth" in Genesis 1:1 is a merism. A merism is a figure of speech in which two or more words are combined to express the entirety of something. For instance, the expression "lock, stock, and barrel" refers

---

[3] For further discussion on this point, see Steven W. Boyd, "Tacking with the Text: The Interconnection of Text, Event, and Time at the *Macro-level*," in *Grappling with the Chronology of the Genesis Flood: Navigating the Flow of Time in Biblical Narrative*, edited by Steven W. Boyd and Andrew A. Snelling (Green Forest, AR: Master Books, 2014), 508–18.

to complete possession of something. The phrase comes from the three basic parts of a rifle: the stock, which holds the other parts and provides something secure for the user to hold onto the rifle, the barrel, through which the projectiles are shot, and the lock (now often called the receiver), which is the mechanism that fires the projectiles. A rifle easily can be disassembled into these three parts, and it is incomplete without all three of these parts.

A merism often contains two contrasting words to convey entirety. An example would be to search high and low for something. It is implied that the search was done at the highest and lowest places, with the implication that everything in between is included. This is the sort of merism that "heavens and earth" is in Genesis 1:1. Even in English today we use this merism, as in "I moved heaven and earth, but I still could not succeed." That expression is used to describe a situation in which every possible approach was tried. Ancient Hebrew lacks a word for "universe." Hence, the only manner to refer to the entirety of physical existence in Hebrew is the use of the merism "heaven and earth."

Second, Genesis 1:1 together with Genesis 2:1–3 functions as an *inclusio*. An inclusio is a literary device that serves to bracket a text with similar material at the beginning and ending of that text. The similar words, phrases, or conceptual material act as a frame or bookends that tie the text together. There are numerous examples of inclusios in the Hebrew Old Testament. One example is Genesis 6:9–10 and Genesis 9:18–19. These verses introduce and conclude the Flood narrative and list the names of Noah's sons. Another example of an inclusio are the first and last verses of Psalm 118, which are word-for-word the same:

> Oh give thanks to the LORD, for he is good;
>   for his steadfast love endures forever!

However, many people reading English translations of the Bible often miss inclusios. There are at least three reasons for this. One reason is that sometimes they are lost in translation. A second reason is that an inclusio is not a common concept in English writing. A third reason is that chapter breaks often intervene between the beginning and end of an inclusio. This is the situation of the inclusio that runs between Genesis 1:1 and Genesis 2:1–3. A good example of inclusio that survives translation well is Psalm 118, where, as we have seen, the first and last verses read identically.

Returning to the creation account, we see an inclusio which involves the introduction of specific works and phrases in Genesis 1:1—בָּרָא ("created"), אֱלֹהִים ("God"), and הַשָּׁמַיִם וְאֵת הָאָרֶץ ("the heavens and the earth")—and their repetition in reverse order in Genesis 2:1–3 (הַשָּׁמַיִם וְהָאָרֶץ in 2:1; אֱלֹהִים, initially, in 2:2; and בָּרָא in 2:3). This inclusio effectively brackets the contents of the creation narrative.

Third, and perhaps most significantly, the grammatical relationship between Genesis 1:1 and 1:2 suggests that Genesis 1:1 contains an instance of introductory encapsulation. The eminent Hebraist Umberto Cassuto noted on Genesis 1:2 that the Hebrew construction וְהָאָרֶץ הָיְתָה תֹהוּ וָבֹהוּ ("Now the earth was formless and void") plainly shows that verse 2 begins a new subject, for he wrote, "It follows, therefore, that the first verse is an *independent sentence* that constitutes *a formal introduction*" (emphasis added).[4] Young concurs with Cassuto's observations, noting there are similarly constructed narratives in the Old Testament that feature summary statements followed by detailed accounts (cf. 1 Kings 18:30ff.). Young phrases his point rather uniquely, saying, "Verse one is a narrative complete in itself."[5] As such, Genesis 1:2 does not follow sequentially what is stated in verse 1 (note the *qatal* verb in 1:2); rather, it is a descriptive statement that represents the state of the world near the beginning of the creative process that is summarized in verse 1. As such, Genesis 1:1 functions to summarize the account of creation to follow, and Genesis 1:2–31 elaborates upon the details of God's creative activity.

Since the creation literature has been dominated by Morris' viewpoint that Genesis 1:1 records the first creative act of Day One, the approach that Genesis 1:1 represents an example of introductory encapsulation is resisted by some recent creationists. There are at least two reasons for this. Genesis 1:1 is foundational to creation, so when one has been accustomed to viewing Genesis 1:1 in a particular manner, it is difficult to conceive of other possibilities. While the caution of this conservative approach is admirable, it is an insufficient reason in itself to oppose the view that Genesis 1:1 contains an instance of introductory encapsulation. It is striking that much scholarship among Christians (who are non-scientists) embraces this concept.

The second reason for resistance to Genesis 1:1 being an instance of introductory encapsulation is the fear that it may lead to belief in billions of

---

[4] Umberto Cassuto, *A Commentary on the Book of Genesis, Part 1: From Adam to Noah*, translated by Israel Abrahams (Jerusalem: The Magnes Press, 1961), 20.

[5] Edward J. Young, *Studies in Genesis One* (Phillipsburg, NJ: P&R Publishing, 1999), 11.

years. However, this is precluded by, among other things, a straightforward reading of Exodus 20:8–11, which states that all of God's creative actions occurred on six normal days. Also, one cannot divorce the introductory statement of Genesis 1:1 from the remainder of the creation account. Thus, one cannot insert billions of years into the first verse or two of Genesis, because the first verse or two do not stand alone. Rather, verse 1 acts as a summary, and the details follow. In the two examples of introductory encapsulation and elaboration repeated above, one about the hypothetical man Bill from today, and the other from Genesis 37:5–7, no one would suggest that considerable time intervened with or after the encapsulatory introduction before the details were related. Instead, everyone readily acknowledges that the events of the encapsulatory introduction and the elaboration must be contemporaneous. To suggest that introductory encapsulation permits for the insertion of great time is to misunderstand introductory encapsulation and elaboration.

Genesis 1:1 being an example of introductory encapsulation presents a possible problem. Someone might object that while the heavens are explicitly detailed as being made on Day Two, there is no explicit statement, apart from Genesis 1:1, of the creation of the earth. Therefore, when did God create the earth? The word *earth* (*'ereṣ*) appears in Genesis 1:1 and Genesis 1:2, but it does not appear again until Genesis 1:9 when God made the dry land to appear and then called it "earth." Genesis 1:9 is in the account of Day Three, so in that sense, the earth did not exist until Day Three, and, indeed, this is the earth that we are familiar with today. Prior to the appearance of dry land on Day Three, what would become the earth is referred to as water (Gen. 1:2). When we ask when God made the earth, it probably is more proper to ask when God created *the material out of which God made the earth*, that is, when did God create the primordial matter that would become the earth as we know it now? To answer that, we need to realize that there is no reason why Genesis 1:1 is merely a merism. Genesis 1:2 describes the unfinished state of the earth (the deep/waters) at the beginning. Since the structure of introductory encapsulation and elaboration does not permit the insertion of additional time, great or small, nor does it permit actions outside of its structure, the initial creation of primordial matter is implied in conjunction with the initial creative acts of Day One. Again, viewing Genesis 1:1 as an example of introductory encapsulation does not permit the addition of billions of years or the creation of things prior to Day One.

God's Work on Day Two

The realization that the mention of the heavens in Genesis 1:1 does not necessarily refer to the creation of the space of the universe frees one up to view the Day Two account differently. Genesis 1:6 declares God's determination to call the *rāqîaʿ* into existence to separate the waters from the waters. This is immediately followed by Genesis 1:7, which states that God made the *rāqîaʿ* to separate the waters above the *rāqîaʿ* from the waters below the *rāqîaʿ*. Finally, in Genesis 1:8, God called the *rāqîaʿ* "heaven" (*šāmayim*), and the Day Two narrative closes. The Hebrew word *šāmayim* refers to things above us, which could include the atmosphere (first heaven) and what we today would call space (the second heaven). There is no reason why both of these could not be included here. The Hebrew Old Testament does not make as sharp a distinction of distance between these two heavens as we might today. However, we can assume that the original readers would have made some distinction based upon observations of objects that appear in heaven. For example, birds, clouds, and precipitation clearly are in the first heaven as they appear quite near the earth, while the sun, moon, and stars (including planets) are in the second heaven.

As an example, consider meteors, which appear about 100 kilometers high in our atmosphere. Being within the upper reaches of the earth's atmosphere, today we would properly consider them to be atmospheric effects, albeit of an astronomical origin. However, the ancient Hebrews would have considered meteors to be stars based upon their appearance (even today meteors commonly are referred to as "shooting stars" or "falling stars"). Hence, the ancients would have considered meteors to be in the second heaven. Artificial satellites did not exist in the ancient world, but within the context of Old Testament language, how would we classify them? They appear as bright stars that move across the sky. From the standpoint of ancient people, including the Hebrews, artificial satellites would be stars. However, they orbit only slightly higher than meteors appear in the atmosphere. The distance between low-earth orbit satellites and the atmosphere is orders of magnitude less than distance between these satellites and the closest astronomical body, the moon. Hence, with regards to distance, even today one could make the case that artificial satellites ought not to be included with astronomical bodies, and so ought not to be in the second heaven.

The point is that the distinction between the first and second heaven is not as clear as some might think, particularly when one views that ancient distinction from a 21st-century perspective. If one were to opine that either only the first heaven or only the second heaven was made on Day Two, then it is not clear where the line of demarcation between the two would have been. However, if both entities were made on Day Two, then this is a moot point.

The word *šāmayim* appears only seven times in Genesis 1. The first three appearances are in Genesis 1:1, 8, and 9, respectively. The first verse is part of the encapsulatory introduction. Verse 8 is God's equation of the *rāqîaʿ* with the *šāmayim*. Verse 9 involves God's command for the waters under the heavens to be gathered into one place and for dry land to appear. Since this immediately follows God's equation of the *rāqîaʿ* and the *šāmayim* and the conclusion of Day Two, it ought to be abundantly clear that the *rāqîaʿ* here ought to be equated with *šāmayim* in verse 9. The four times that *šāmayim* is used in the remainder of Genesis 1, it always appears in construct with the word *rāqîaʿ*, as it is translated "firmament of heaven" in the King James Version. Three of these uses are in the context of the Day Four account (verses 14, 15, and 17), with the fourth appearance in the Day Five account (verse 20). The implication seems to be, lest there be any confusion, that this entity mentioned is the same thing that God made on Day Two. Within the context of the Day Four narrative, this "firmament of heaven" is where God placed the luminaries—the sun, moon, and stars. In the Day Five account, the birds are said to fly "across [*or* upon] the expanse of the heavens." The construction in Genesis 1:20 is different from the other appearances of the phrase *firmament of heaven*, and it is difficult to translate. There is a distinction from where the stars are, suggesting that the birds merely fly across the interface of the firmament of heaven. These considerations and others suggest that the *rāqîaʿ* is closest to what we would call the sky. In this respect, the atmosphere, and especially the lower portions of the atmosphere, may be considered the near surface of the *rāqîaʿ*.

Besides being based upon a clear reading of the Genesis 1 creation account, this understanding of the *rāqîaʿ* nicely incorporates the Old Testament verses that speak of the heavens being stretched or spread out—as in Job 9:8; Psalm 104:2; Isaiah 40:22; 42:5; 44:24; 45:12; 48:13; 51:13; Jeremiah 10:12; 51:15; and Zechariah 12:1. Besides these 11 verses, there are a few other verses that might qualify for inclusion, but they are not so clearly worded, so they were omitted from the list above. One of those verses worthy of note is Job 37:18, which says that God has spread out the sky. The word "sky" rather than "heaven" appears here, because the noun שְׁחָקִים (sᵉḥāqîm) is used rather than šāmayim. Sᵉḥāqîm literally means "clouds," and so is associated with the sky. It is not necessarily equivalent to šāmayim, but because clouds are in the atmospheric heaven there is some overlap in the semantic range of the two words. Interestingly, the Hebrew verb in Job 37:18 that is translated "spread" is רָקַע (rāqaʿ), the verb from which the noun rāqîaʿ comes. By contrast, the verb in each of the 11 verses listed above is נָטָה, meaning "to stretch" or "to spread out." However, it also can mean "to bend." This verb appears in 2 Samuel 22:10; Psalm 18:9; and Psalm 144:5, which says that God bent, or bowed, the heavens—so these verses say much the same thing and could be included as well, though they read differently in English.

Admittedly, the Hebrew verbs used to describe the spreading or stretching of the heavens do not appear in the Day Two account. The verb in Genesis 1:7 is עָשָׂה, which means "to do, make, or perform," and is commonly used in the creation account in reference to God's creative action. However, the word rāqîaʿ derives from the word rāqaʿ, a verb meaning "to beat, to stamp out." This is the sort of action that one might do with a malleable metal, such as gold. Through mechanical means, craftsmen can beat, stamp, or otherwise spread gold into very thin sheets, often for the purpose of inlaying objects. In recent years, it has become common to view the stretching of the heavens as referring to the expansion of the universe. However, as universal expansion was not discovered by Edwin Hubble until 1929, this would not have been how the text would have been understood prior to the 20th century. Certainly, those who wrote about the stretching of the heavens or those who first read or heard it must have had some understanding of what this meant. In each case where the stretching of the heavens is mentioned in the Old Testament, the context is within the discussion of the sovereignty and omnipotence of God based upon His role as Creator. Therefore, one ought to look into the creation account for the meaning of these passages. Since Genesis 1:8 equates šāmayim with rāqîaʿ, and we know the verb from which rāqîaʿ comes means "to beat" or "to spread out," the best fit for understanding the stretching of the heavens is with what God did on Day Two. The "stretching" out of the heavens thus refers to a past event, not an ongoing phenomenon.

Hebrew verbs do not innately possess tense as English verbs do, so properly translating them occasionally can be tricky. However, in a number of the 11 verses that mention the stretching of the heavens, the action is associated with creation, which is finished (Gen. 2:2). For instance, consider Isaiah 42:5. The verse begins,

> Thus says God, the LORD,
>   who created the heavens and stretched them out . . .

Notice the parallelism between "created the heavens" and "stretched them out." Since the creation of the heavens is a past event, the parallel makes no sense if the stretching out is ongoing. Similar parallel structures tying the stretching of the heavens to the (past) creation process exist in Isaiah 44:24; 45:12; 48:13; Jeremiah 10:12; 51:15; and Zechariah 12:1. It is difficult to reconcile the ongoing expansion of the universe with verses describing the stretching of the heavens at the time of creation.

Especially noteworthy is another example of parallelism found in Psalm 19:1. This verse has tremendous bearing on a correct, biblical understanding of the *rāqîaʿ*. The verse reads,

> The heavens declare the glory of God,
>   and the sky above proclaims His handiwork.

Here the Hebrew word *rāqîaʿ* is rendered "sky." This verse says the same thing two different ways. However, the parallelism works only if the two subjects, *šāmayim* and *rāqîaʿ*, are equivalent on a conceptual level (i.e., they refer essentially to the same thing). These two entities are exactly equated in Genesis 1:8, so they are the same. Therefore, if the *rāqîaʿ* made on Day Two is the earth's atmosphere, then the subject of Psalm 19:1 is the earth's atmosphere. No one believes this to be the case, for Psalm 19:1 is universally accepted as referring to the celestial heavens. While Psalm 19 does not specifically mention stars, it does mention the sun in verse 4, and the sun is further discussed in verses 5 and 6. Clearly, God made the sun on Day Four and placed it in the firmament, or expanse, of heaven. This is not the earth's atmosphere.

The Waters Above the Heavenly Expanse

God made the *rāqîaʿ* to separate the waters above it from the waters below it. If the *rāqîaʿ* can best be identified principally as what we call space today, then there are three startling conclusions. First, the universe is bound, or has an edge. While this possibility is permitted within the physics of space and time as we now understand it, this position decidedly is unpopular among cosmologists. If the universe is unbound, then the universe either can be finite or infinite in size. If finite, then the universe has curvature so that space closes back on itself so that there is no boundary. Outside of the creation literature, very little work has been done on cosmological models that are bound.

Second, since the *rāqîaʿ* was spread out from the waters below the *rāqîaʿ*, and the earth formed out of those waters, unless this spreading was asymmetrical, then the earth must be at, or at least near, the center of the universe. In developing their cosmologies, D. Russell Humphreys and John Hartnett have suggested that the earth is near the center of the universe, albeit by slightly different means. Among non-biblical cosmologies, this is to be resisted more strenuously than a bound universe. The reason is that this runs counter to the Copernican principle, a commonly-held belief today that asserts that the earth is in no particularly significant location. Most cosmologies today deny that the universe has a center, opting for either an infinite universe or an unbound finite universe. In either case, the universe has no center. There is no way at this time observationally to determine if either of these views is correct. Even if the universe had a center, the probability of the earth being near that center in the vast universe is vanishingly small. Hence if it turned out that the universe had a center and the earth were near it, that highly improbable location would imply design and a Creator.

Third, the Bible implies that the boundary of the universe is accompanied by water. Unlike what the canopy model proposes, the waters above the *rāqîaʿ* did not condense at the time of the Flood, and so still ought to be beyond the *rāqîaʿ*. This is borne out by Psalm 148:4, which speaks of waters above the heavens still being there. We do not know who wrote Psalm 148 or when he wrote it, but it almost certainly was long after the Flood. That is to say, in the post-Flood world, the universe is still surrounded by water.

What form might this water at the edge of the universe be in? Some might wish to have this water in a solid or gaseous form, as opposed to liquid. However, the Hebrew word for water, מַיִם, is used expressly for liquid water. If ice were intended, the word would be קֶרַח. If gaseous water were meant, we might expect to see אֵד, or perhaps הֶבֶל. Therefore, the water above the *rāqîaʿ* on Day Two must have been liquid. Furthermore, Psalm 148:4 suggests that at least at the time that Psalm 148 was written, the water above still was in a liquid form. Some might object that the conditions in space are such that liquid water could not remain in that state, but instead must have condensed into ice or evaporated into gas. However, we know nothing of the physical conditions at the edge of the universe. Indeed, the edge of the universe is a difficult concept to grasp physically. It may be that God has imposed conditions at the edge of the universe so that the water there remains as a liquid. Or perhaps not. While I prefer liquid water at the edge of the universe, I shall now consider the implications of water at the edge of the universe, not only as a liquid, but also as a gas or solid.

All baryonic matter (such as water) must radiate, if its temperature is above absolute zero. We have never observed, nor can we conceive of matter, with absolutely no temperature, so the assumption that the water at the edge of the universe has temperature seems warranted. Solids, liquids, and gases at high pressure radiate a blackbody spectrum that is a function of temperature.[6] A question arises as to whether the water at the edge of the universe is optically thick.[7] I will assume here that it is, thus ensuring a clean blackbody curve. If the water at the edge of the universe is a gas at low pressure, it will produce an emission spectrum,[8] which will be a function of its temperature. At any rate, the spectrum of a low pressure gas will be dramatically different from the spectrum of the other possibilities. Assume that the temperature of the water is 300 K (Kelvin), close to room temperature. This water must lie beyond the most distant galaxies or other objects in the universe. Observationally, we know that there is a direct relationship between distance and redshift (the

---

[6] In physics, an ideal blackbody absorbs all radiation that falls on it. No real object is an ideal blackbody, but many objects approximate ideal blackbody behavior. Since no light is reflected, a blackbody at or near room temperature appears perfectly black, hence the name. However, if a blackbody is sufficiently hot, it will radiate energy in the part of the spectrum that the eye can see, so blackbodies do not always appear black. Not only do blackbodies absorb radiation perfectly, they are also perfect emitters of radiation. The spectrum emitted by a blackbody has a very characteristic shape that is dependent upon temperature. A blackbody spectrum emits in a broad range of wavelengths. This often is called a continuous spectrum.

[7] This is a very technical term. For purposes here, suffice it to say that being optically thick means that no significant amount of light could penetrate the layer. Given their white color, ice and droplets of water (in the form of clouds) are optically thick. While a thin layer of liquid water is optically thin in the visible part of the spectrum, it is optically thick in the infrared, which is significant here.

[8] An emission spectrum is very different from a blackbody spectrum. An emission spectrum emits radiation only at very narrow wavelengths rather than broadly at a wide range of wavelengths.

Hubble relation).[9] Therefore, the spectrum of the water must be redshifted by a factor greater than the largest observed redshift. Currently, the record for greatest redshift is on the order of ten. Assuming this value as the redshift of the spectrum given off by the water at the edge of the universe results in a blackbody spectrum of 30 K (this is −410°F). Keep in mind that this is just an estimate, not a prediction.

What do we observe? The universe appears to be bathed in a radiation field called the cosmic microwave background (CMB). The currently measured temperature of the CMB is 2.725 K. Since 1965, the CMB has been interpreted as the best evidence for the big bang model. Presumably, the CMB emanates from a time nearly 400,000 years after the big bang when the universe was sufficiently hot and dense enough to be opaque. According to the big bang model, once the universe had expanded and cooled sufficiently, the universe became transparent, and matter and photons decoupled for the first time, thus permitting the light from the opaque gas at the time of decoupling to reach us. After traveling over billions of light years, the blackbody spectrum of the opaque gas has been redshifted by a factor of about a thousand, thus cooling the blackbody curve of the gas from about 3,000 K to about 3 K.

One problem for recent creationists who reject the big bang model is the lack of explanation for the CMB. However, if water truly is at the edge of the universe as Genesis 1:6–8 suggests, then we ought to expect that the universe is surrounded by water, which ought to radiate. Assuming cosmological redshift, regardless of its cause, the radiation from this water ought to be a cool blackbody, which is what we observe. It was possible that between 1929, when Edwin Hubble discovered the expansion of the universe, and 1965 someone could have predicted the CMB, if they had taken Genesis 1:6–8 seriously.

The Planet Earth

Let us turn our attention to the earth. The Hebrew word 'ereṣ appears more than 2,500 times in the Old Testament. English translations render 'ereṣ a number of ways, with "earth," "country," "ground," and "world" being the most common

---

[9] In 1912, Vesto Slipher showed that the radiation from most galaxies is shifted toward longer wavelengths. Since red is on the longer wavelength limit of what the eye can see, we call this redshift. Building on this work, in 1929, Edwin Hubble showed that the redshifts of galaxies are related to their distances. This is the Hubble relation. Turning this around, we can use the Hubble relation to infer distance. The most straightforward interpretation of the empirical Hubble relation is that the universe is expanding.

renderings. These uses run parallel to our modern English word *earth*. We use this word to describe our planet, but also to refer to soil or to real estate as a particular part of this planet. As happens so many times in the Bible, the context normally determines exactly what meaning is intended. As previously discussed, *'ereṣ*, along with and preceded by the Hebrew word *šāmayim*, and translated as "heaven and earth," constitute a merism in Genesis 1:1, referring to the totality of creation.

As mentioned earlier, the likely meaning of *'ereṣ* in Genesis 1:2 is the material from which God made our planet earth as we now know it. Genesis 1:2 describes this planet as being "formless and empty" (תֹהוּ וָבֹהוּ). Some have seen in this verse a catastrophe that befell the earth and caused it to become formless and empty (the gap theory), but this is not supported by good exegesis. Instead, verse 2 describes the initial condition of the matter making up the earth, before any shaping or sculpting. It is sort of like a lump of clay or a piece of stone that an artist acquires but has not yet begun to work on.

Two other points stand out in Genesis 1:2. First, immediately after the earth is described as formless and empty, the text says that darkness was over the face, or surface, of the deep. The Hebrew word for "deep" is תְהוֹם (*tᵉhôm*). The word *tᵉhôm* appears 36 times in the Old Testament. Each time this word is used it refers to depths of water, such as the ocean or some other large body of water. The Septuagint translated *tᵉhôm* as ἄβυσσος, from which we get the English phrase "watery abyss." Second, the verse says that the Spirit of God was moving over the surface of the waters. We see that the early, unfinished earth either consisted of water or was covered with considerable water. The Apostle Peter in his second epistle (2 Pet. 3:5) makes reference to this in that the earth was formed both out of water and by water. In fact, words referring to water appear several times in the Genesis 1 creation account, so water was a very important part of the creation. As for the Spirit of God moving upon the face of the waters, this might refer to God's attention turning to shaping and preparing the earth for living things and, most importantly, man.

Light Created on Day One

Verse 3 commences the first of the six statements beginning with the words, "And God said, 'Let there be....'" The power of God's command is very strong. Psalm 33:6 and 148:5 echo that the creation happened upon God's command. On this first day, God commanded the creation of light. Verse 4 tells us that God saw that the light was good, and that he separated the light from the dark.

Verse 5 records that God called the light "day," and the dark He called "night." The naming of night and day is immediately followed by the statement concluding the first day of creation. This is one of the best evidences that the intended meaning of the word *day* in the creation account of Genesis 1 is a normal day. In attempting to accommodate vast periods of time into the creation account, many Christians argue that the Hebrew word for day, יוֹם (*yôm*), can mean an indefinite period of time, and thus conclude that the six days of the Creation Week were time periods. We call this the day-age theory. That stated, there is nothing mysterious about the word *yôm*. *Yôm* does have four distinct meanings, a 24-hour light/dark cycle, the light portion of that cycle, an indefinite period, and a time appointed for some purpose. The English word *day* has those same meanings, as does the word for "day" in most languages. The important question is not what possible meaning a particular word may have, but what is the most likely meaning of that word within the context of its usage.

There are a number of contextual reasons why a normal day is intended in Genesis 1 and none that actually point to a long period of time. One of the contextual reasons is that God has just defined what a day is in the context of light and dark, and then He immediately declares the conclusion of Day One of the Creation Week. Another argument for the days being normal days is the description of the days having an "evening" and a "morning." This use normally is associated with normal days, and it amounts to metaphorical use to claim that these terms refer to the beginning and ending of a long time period. Another argument is the numbering of the days in the Genesis 1 creation account. In English, this usually reads as "the first day," "the second day," and so forth. These are ordinal numbers, and Days Two through Six are expressed as ordinal numbers in the Hebrew. However, in Hebrew the first day is an exception, for it is expressed as a cardinal number (אֶחָד; Gen. 1:5) and reads as "one day." The point, it seems, is that Moses in Genesis 1:5 is deliberately indicating that a "day" is a period of time that is marked by an "evening" and a "morning." Moreover, in nearly every case in the Old Testament, when days are numbered, it refers to normal days. This convention also is followed in English—if an account numbers days, it refers to a normal day. Another argument comes from Exodus 20:11 where God gave instruction concerning the Sabbath Day (certainly understood as a normal day) by comparing it, the seventh day, to God's resting from His work on Day Seven of the Creation

Week. This is just a cursory discussion of the reasons why the days of the Creation Week were normal days—there are more complete discussions elsewhere.[10]

What was the source of the light? Most of the light on the earth now comes from the sun, but the sun could not have been the source of the light for the first three days, because God did not make the sun until Day Four. Skeptics frequently ridicule the Bible on this point, arguing that ignorant and foolish people wrote the Bible since they didn't see a problem with the fact that the sun didn't exist for the first three days of creation. However, the creation account doesn't identify the source of the light for the first three days, so we don't know what the source was, but we can be sure that it was not the sun. Obviously, the light source was replaced by the sun on Day Four.

There has been speculation about the light source for the first three days, and as long as we understand that this is speculation, there is no real harm in doing so. The most common speculation is that God was the source of the light. Second Corinthians 4:6 paraphrases the creation of light on Day One and goes on to draw a parallel to God shining light in our hearts "to give the light of the knowledge of the glory of God in the face of Jesus Christ." Revelation 21:22–23 states that the New Jerusalem in the eternal state will experience no night, and that the city will need neither the sun nor the moon to shine on it, for the glory of God will illuminate it, and its lamp will be the Lamb. There are parallels between the original edenic state and the coming eternal state, and so the source of the light the first three days may have been the radiance of God Himself as it will be in the world to come. Notice that Revelation 21 does not state that there will be no sun or moon in the eternal state; rather, it states that there will be no need for their light in the New Jerusalem. Revelation 21:1 tells us that there will be a new heaven and a new earth. It is possible that the new heaven will have a sun and a moon, but we will not require their light to complete our tasks.

We ought to make one other observation here. The account of each of the six days of the Creation Week concludes with a statement that sounds a bit peculiar to our ears today. The Day One account ends with the statement, "And the evening and the morning were the first day," with each subsequent day's account using an incremented number. Today we think of morning

---

[10] For instance, see Andrew E. Steinman, "אחד as an Ordinal Number and the Meaning of Genesis 1:5," *Journal of the Evangelical Theological Society* 45, no. 4 (2002): 577–84.

coming before evening, and so this usage sounds backwards to us. However, in traditional Jewish reckoning, the day begins and ends at sunset. For instance, the Sabbath begins at sunset of what we would call Friday evening and ends at sunset of what we would call Saturday evening. So listing evening before morning makes more sense in this manner of reckoning days.

Conclusion

It is a common belief among recent creationists that Genesis 1:1 describes the first creative acts of Day One. Within that view, the creation of the heavens in Genesis 1:1 is assumed to be the creation of what we would call space today. If this is correct, then the entity made on Day Two, the *rāqîaʿ*, must be something else. But there are good reasons to believe that the *rāqîaʿ* is what we today call space. An easy solution to this dilemma is that Genesis 1:1 contains an example of introductory encapsulation and that Genesis 1:6–8 ought to be understood primarily in terms of the creation of what we now call space (or sky). Some recent creationists may object on the grounds that Genesis 1:1 as an instance of introductory encapsulation is a retreat, or that some people with a belief in billions of years also believe this. However, many of those who believe in billions of years also believe in the cardinal doctrines of Christianity, such as the deity, Virgin Birth, and bodily Resurrection of the Lord Jesus Christ. By that reasoning, we ought to reject these, because some people who believe in billions of years also believe these cardinal doctrines. As for fear of changing one's mind about such a matter, we ought not to develop our theology or our creation model motivated by fear. To the contrary, our commitment must be to the integrity of Scripture and to what the Bible actually says.

How one interprets Genesis 1:1 directly affects how one interprets Genesis 1:6–8. If one gets Genesis 1:6–8 wrong, it will have little, if any, impact on a biblical model of geology. If one gets Genesis 1:6–8 wrong, it will have little, if any, impact on a biblical model of biology. However, it one gets Genesis 1:6–8 wrong, then there is little hope of developing a successful biblical model of astronomy.

CHAPTER 3

Biblical Teachings about the
Sun, the Moon, and the Stars

The Day Four Account of Creation

The Day Four account found in Genesis 1:14–19 describes the creation of the astronomical bodies. As with five of the other six days of the Creation Week, the account begins with, "And God said...." On this day, God commanded that there be lights in the expanse of heaven. Verse 14 goes on to list purposes for the lights: to divide the day from the night, and to be for signs, and for seasons, and for days and years. Verse 15 gives an additional purpose for them, to be lights in the expanse of heaven and to give light on the earth, a purpose repeated in verse 17. Verse 16 says that God made two great lights, the greater light to rule the day, and the lesser light to rule the night. Verse 18 restates this purpose and reports that God saw that this was good.

Notice that God here did not use the normal Hebrew words to identify the sun and the moon. Instead, He chose to refer to them as "great lights." Even in verse 16, He chose to describe them as the "greater light" and the "lesser light." These obviously are the sun and moon, respectively, so why did God not use the usual words for the sun and the moon? The Hebrew word for "sun," שֶׁמֶשׁ (šemeš), first appears in Genesis 15:12, and a word for the moon, יָרֵחַ (yārēaḥ), doesn't occur until Genesis 37:9. We don't know for certain why God chose not to use the normal words for the sun and the moon in the creation account. However, the most probable explanation is that the sun and moon were regarded as deities by most people in the ancient Near East, and that by omitting the names of the sun and moon, the text intends to convey the theologically significant point that the sun and moon not only are not objects to be worshiped, but also that they are not personal entities in any sense.[1] Rather, they are creations of God and made to serve man. The only time that the stars are explicitly mentioned in the creation account is at the end of verse 16.

The Purposes of the Heavenly Bodies

The Day Four account gives several purposes of the astronomical bodies. They are:

1. To divide the day from the night
2. To be for signs, for seasons, and for days and years
3. To be lights in the expanse of heaven and to give light on the earth

Additionally, the sun and moon were given a special purpose:

4. To rule over the day and to rule over the night.

The first, third, and fourth purposes are directly related to one another. The sun, moon, and stars are lights in space, and they provide light on earth. Of course, the sun is the dominant source of light on the earth, and its light very clearly delineates where it is day, and the absence of its light delineates where it is night, thus dividing day and night. In the sense that the sun is the most noticeable, and sometimes the only noticeable, object in the sky during the day, it certainly rules over the day. Likewise, when the moon is visible at night, its light dominates and hence rules over the night. However, Psalm 136:9 tells us that both the moon and stars rule by night. What does it mean to rule over the day or night? Certainly being the dominant light is a part of that, but the light itself can be useful and hence may be part of that ruling as well. The light of the sun allows us to see well during the day, but a bright moon can allow people to see well at night. For example, Passover always is at full moon, because the first Passover was at full moon of the first month of what became the ceremonial year. Exodus 12 records that it was at night when Pharaoh told the Hebrews to leave at once. Thus, their journey began at night, and the full moon provided extra light for them to travel. Even if the moon is not up, the light of stars in a very dark location allows people to see well enough to get around. It is more difficult, but if there are no overhead obstructions, one can see well enough just by starlight to get around on a dark, moonless night.

---

[1] Notably, the Hebrew word *šemeš* is associated with the Semitic name of a pagan sun deity, so its absence in the creation account may have been not to even dignify its use. However, the Hebrew world for moon, *yārēaḥ*, is unconnected with a lunar deity name.

What does part of the second purpose, to be "for seasons," refer to? "Seasons" can have two distinct meanings. One meaning for seasons is reference to the manner in which weather changes throughout the year. In much of the world the most obvious change is in temperature. But in tropical or Mediterranean climates another obvious change is the amount of rainfall. In much of the temperate world, we recognize four seasons, winter, spring, summer, and autumn. However, in some parts of the world there may be only two or three noticeable seasons. For instance, in some tropical regions, people recognize dry and rainy seasons. The seasons are the result of the earth's axial tilt and the earth's orbit around the sun. Throughout history, the vast majority of people were directly involved with agriculture. Only in very modern times and then only in industrialized societies have people had the luxury of not needing to grow their own food to survive. Most people reading this book likely aren't directly tied to agriculture, and so we often miss the profound importance of the proper anticipation of seasons. It is easy to succumb to the temptation to plant crops during a warm spell in early spring, but experience has shown that crops planted too early are at great risk to damage from cold weather. It is critical that a good calendar reveals when the proper time to plant is, and a good calendar relies upon carefully noting the changing position of the sun with respect to the stars. The stars that we see change throughout the year, because of our revolution around the sun. In the Northern Hemisphere, Orion is a winter constellation, but Scorpius is a summer constellation. In this sense, stars indicate the seasons.

While this is the overwhelming meaning of the word *season* today, this hasn't always been the understood meaning. The original meaning was more broadly in reference to a period of time or to a time appointed for some purpose. An example of this usage is Hebrews 11:25, which speaks of Moses choosing to suffer affliction with his people rather than enjoying sin for a season. Obviously, one of our climate-related seasons is not the intended meaning here. Today we have various hunting and athletic seasons. Deer season is a period of time when deer hunting is legal. Baseball season begins in the spring and continues until early autumn. We also speak of "the Christmas season" or "the holiday season," the period of time just prior to Christmas and extending just past New Year's Day. Many ancient cultures had similar observances, or seasons. For instance, the Hebrews were instructed to observe Passover, Pentecost, and Yom Kippur with the related Feast of Tabernacles.

Later on, the Hebrews added other observances, such as Hanukkah. Each of these observances stretched out over several days, and so were a sort of season. In fact, the Hebrew word translated "season" (מוֹעֵד) in Genesis 1:14 literally means "appointed time," and elsewhere specifically refers to some of these observances and more properly could be translated "festival" (e.g., Isa. 33:20). Since we believe that Moses wrote the Pentateuch (including the creation account and the instruction of observance of Passover and other festivals), the Hebrews of that time likely would have made the association between the seasons mentioned in Genesis 1 and the festivals. To know when each of these celebrations was to be observed required an accurate calendar. Again, an accurate calendar required observation of the sun and the moon with respect to the stars. Thus, throughout history one of the primary functions of astronomy was to produce and maintain calendars. Chapter 4 discusses calendars in more detail as they relate to the purpose of the heavenly bodies.

Time of Day

In Chapter 4 we will discuss the natural units of time, the year, month, and day, as well as the unnatural unit of time, the week. There is a need to further to subdivide the day. We artificially subdivide the day into hours, hours into minutes, and minutes into seconds. These divisions of time are base 6 and 60, as is the division of a circle into 360°. Historians think that base 6 and 60 measurements originated in ancient Babylon. The ancient Babylonians placed religious significance on base 6, but it's not clear if the religious significance or the practice of using base 6 came first. (This may be the significance of the number of the beast, 666, in Rev. 13:18, for Babylon figures greatly in the next five chapters of Revelation.)

The division of time into hours, minutes, and seconds is arbitrary, and almost certainly was not developed immediately after creation, but was introduced much later. It has no biblical basis, but that does not necessarily mean that its use is anti-biblical. In our modern world we often think that base 10 is superior. This works well with computation of large numbers, but not so well with small quantities. When working with small quantities, fractions are advantageous. This is the reason why many old non-metric systems of measure use base two and three. Fractions work well with two and three, but not with ten. Since six is divisible by both two and three, both base 12 and 60 are good choices for subdividing the day. The word *hour* occurs only five times in the Old Testament, with all five in the book of Daniel (interestingly, much of the setting of the book of Daniel is Babylon). The Aramaic word used there is שָׁעָה, and it refers to a brief time or moment, not an hour as we mean it today.

Sunrise and sunset are frequently mentioned in the Bible, and as such they amount to measurements of time. However, the first mention of measuring any other time of day is the sign that Isaiah the prophet offered to King Hezekiah in 2 Kings 20:8–11. Isaiah gave Hezekiah a choice between a shadow moving ten steps forwards or back, and Hezekiah chose the latter, reasoning that eventually the shadow would move forwards, but it would require a more impressive miracle to move the shadow backwards. Apparently, a shadow of some object falling on a stairway functioned as a type of sundial. On a sundial, the object whose shadow measures the passage of time is called a gnomon. We don't know anything about the construction of Hezekiah's sundial or what its gnomon was. It could be that the stairway and gnomon were specifically designed to act as a sundial, or people may have realized after its construction that the stairway and some other nearby tall object acted as a sundial. Some people argue that the device here was an actual sundial and not a stairway at all and that since 10° is the equivalent of 40 minutes, the sun rapidly moved backwards 40 minutes. The reason for this belief is that the King James Version uses the word *degrees* and even states in 2 Kings 20:11 that it was a sundial. The Hebrew word that the King James Version translates as "degrees" and "sundial" here is מַעֲלָה (*ma'ălâ*). This word means "step," which is how the word is frequently translated in the Old Testament. However, the only other places the King James Version translates *ma'ălâ* as "degrees" are in the accounts of Hezekiah in 2 Kings 20:8–11 and in Isaiah 38:8, as well as in the introduction of each of Psalms 120–134 as "A Song of Degrees" (in other versions typically translated as "A Song of Ascents"). Obviously, the use of *ma'ălâ* does not refer to degrees as an angular measurement in the introductions to those Psalms. Nor does *ma'ălâ* refer to degrees as angular measurement in the other Old Testament appearances outside of 2 Kings 20 and Isaiah 38. Given that when describing physical objects *ma'ălâ* normally means a step or steps, most commentators and modern translations go with "steps" and "stairway" rather than "degrees" or "sundial" as the King James Version does. So we don't know how much the sun actually reversed direction at this time. Chapter 6 discusses the miracle of Hezekiah's sundial, or steps, in more detail.

The word *hour* occurs more than 80 times in the New Testament. The Greek word used is *hora*, which can mean a moment, a season, a day, or the same thing that we mean today, one-twelfth of daylight. The context normally makes it clear which meaning is intended. By New Testament times, the Jews had adopted subdivision into 12 hours of day and 12 hours of night. The hours of day began at sunrise, and the hours of night began at sunset. Therefore, the third hour is 9:00 a.m. or 9:00 p.m., the sixth hour is noon or midnight, and the ninth hour is 3:00 a.m. or 3:00 p.m. The hours were evenly spaced throughout the day and night. Since the days are longer in summer than in winter, a daytime hour in summer was longer than a daytime hour in winter. Conversely, hours at night in the summer were shorter than hours at night in winter. Since there were no accurate mechanical or electronic clocks as we have today, this arrangement worked well in the ancient world, though we would find it peculiar.

The only day of the week that has a name in the Bible is Saturday (Sabbath). All the other days, when referred to at all, are simply called the first day of the week and so forth. This is not surprising, since our names of the days are of more recent origin and come from the names of pagan gods (some Nordic and some Roman). The names of the days in the Romance languages come from gods in the Roman pantheon. This apparently was a common practice in the ancient world, and so the Bible probably avoided using those pagan names for days to avoid any taint of paganism. While our weekday names have pagan roots, the practice of those pagan religions fell away so long ago that no one seriously argues that their continued use constitutes a sort of endorsement of paganism. (Chapter 4 further discusses our modern names for the days of the week.)

What of the remaining purpose for astronomical bodies, to be for "signs"? This one is the most nebulous of the purposes. There are several biblical answers for what these signs may be. First, in Matthew 16:1–4, the Pharisees asked Jesus for a sign. He responded by quoting from some of their own teachings about the sign of a red appearance in the sky to forecast weather. This is similar to our adage, "Red sky at night—sailor's delight; red sky in morning—sailor take warning." Jesus went on to chide the Pharisees for not recognizing the signs of the times. Thus, in context, the people understood that this form of weather forecasting using the sky was a kind of sign. Second, as Psalm 8, Psalm 19, and Romans 1:18–20 tell us, God's existence is revealed through the heavens. Third, the star that led the magi to the infant Jesus (Matt. 2:1–2, 9–10) was undoubtedly a sign from heaven. Fourth, there will be signs in heaven that reveal God's wrath (Isa. 13:9–13; Joel 2:30–31; Matt. 24:29–31; Mark 13:24–27; Luke 21:25–28; Rev. 6:12–17). Chapter 9 further discusses prophetic signs in the heavens. Thus, in biblical passages we have at least four types of possible signs in heaven and heavenly bodies that conform to the God-ordained purpose for them.

The Stars

The only time that the stars are explicitly mentioned in the creation account is at the end of verse 16, where it states, "He made the stars also." The words "He made" are in italics in some English translations (such as the King James Version), indicating that those words are not in the Hebrew, so the account there actually reads, "the stars also." (Note, however, that the italicized words are warranted, since the untranslated direct object marker אֵת in the Hebrew text ties the mention of the stars to the verb "made" at the beginning of the verse.) It would appear at first glance that stars are a mere afterthought, but they are included in the lights commanded in verse 16. This must be the case, for the two great lights are distinguished from the other lights in the expanse of heaven, and the only other lights in the expanse of heaven are the stars. Besides, some of the functions of the two greater lights can only be fulfilled by reckoning their motion with respect to some reference, and the stars provide that reference, as we shall soon see.

Biblically, what is a star? The Hebrew word for star is כּוֹכָב (kôkab), and means "burning" or "blazing." Interestingly, Kochab is the name of Beta Ursae Minoris, the second brightest star in the Little Bear (or the Little Dipper). The word kôkab appears 37 times in the Old Testament. In all but one case it is translated as "star." The one exception is Isaiah 47:13, where it is translated "stargazers." This translation is warranted, as here kôkab appears in construct with the plural form of hāzāh, a word which means "seer." The Greek word for "star" is ἀστήρ (aster), a word from which we get the words astronomy, astrology, and asteroid. Aster appears 24 times in the New Testament, where it is translated "star" each time. To the ancients, a star is any bright, point-like object in the sky, usually visible at night. This obviously would include the objects that we know today as stars, but it would include other objects that we don't currently recognize as stars. For instance, the five naked-eye planets, Mercury, Venus, Mars, Jupiter, and Saturn, appear as bright stars, so in the ancient sense, they were considered stars. Stars move, but they move so slowly that the stars remain fixed with respect to one another over many human lifetimes. However, the planets appear to move with respect to the background stars, so the ancients called these "wandering stars." The Greek word for wandering is πλανήτης (*planetes*), from which we get the word *planet*. In fact, the phrase, ἀστέρες πλανῆται, appears in Jude 13, which is commonly translated "wandering stars." Therefore, the naked-eye planets are stars in the biblical sense

There are many celestial objects too faint to be seen without a telescope, so they weren't known to the ancients. However, since they mostly differ only in degree from objects visible to the eyes alone, it follows that these fainter objects are stars in the biblical and ancient sense as well. Obviously, this includes the stars too faint to be glimpsed by the eye alone. The two planets too dim to be seen with the naked eye (Uranus and Neptune) are also stars in the biblical sense. Asteroids are generally too faint to be seen with the naked eye, but with a telescope they have an appearance very similar to stars, so they, too, are stars in the biblical sense. We could reach the same conclusion about the satellites, or moons, of the planets. Today, artificial satellites orbiting the earth may be visible as star-like objects noticeably moving against the background of stars, so they can be considered stars too. A meteor is a piece of debris that burns up as it enters the upper atmosphere. A meteor appears as a brief, bright flash that resembles a star, so even today people call these shooting stars or falling stars. To the ancients, these would be stars. On rare occasions a star appears where none had been visible and then fades from view over days or months. The Chinese called these "guest stars," but in the West they generally were called "new stars." The Latin word for new is *nova*, so we call such a star a nova (plural, novae). These too are stars in the biblical sense. Today we believe that a nova is an eruption of material on the surface of a particular type of star in a close binary star system. We also recognize a much brighter and rarer event called a supernova. Supernovae can have several different causes, but they all are explosions of stars.

The ancients viewed comets as stars too, though they are extended and fuzzy. The ancients called these "hairy stars." The word *comet* comes from the Latin word *coma*, which means "hair." The ancients noticed a few other fuzzy objects amongst the stars, but unlike comets, they did not move with respect to the stars. The Greeks considered these to be cloudy stars, and the Greek word for cloud, *nebula*, has stuck for these. With the invention of the telescope, some of the nebulae (plural form) were revealed to be clusters of stars too faint to be seen individually. Other nebulae turned out to be vast, distant groupings of stars that we now call "galaxies." Still other nebulae turned out to be true clouds of gas and dust in interstellar space. In addition, the telescope allowed the discovery of many more nebulae too faint to be seen with the naked eye. All of these newly discovered nebulae are true clouds, star clusters, or galaxies. In the ancient sense, all of these objects are "stars," and so biblically we can treat all astronomical bodies, other than the sun and the moon, as stars. Hence, it would seem that the mention of God making the stars on Day Four applies to all astronomical bodies. Therefore, we conclude that God made all astronomical bodies on Day Four.

In modern astronomy we think of the sun as being a star. However, the Bible never refers to the sun in this way. That does not mean that it is wrong to classify the sun as a star in the modern sense. Rather, this underscores the fact that much of astronomy in the Bible is phenomenological. That is, it is based upon how things appear to us. Even today, we generally do not think of the sun in the same context that we think of stars. The obvious reason for this is how incredibly bright the sun appears as compared to other stars, due to how much closer to us the sun is. This makes the sun special to us, for it provides the heat and light that makes life possible.

However, might there be other ways in which the sun is special? As it turns out, there are at least two other ways that the sun stands out. First, the sun has far less lithium than most stars do. Even compared to stars that astronomers consider to be similar to the sun, the sun's lithium is extremely low. It is not clear what this might mean, or whether it is at all significant. Second, astronomers have searched for solar analogues, stars that resemble the sun enough to almost be its twin. Such stars are very rare. That vast majority of stars that resemble the sun vary in brightness. With all the concern with climate change today, can you imagine how much the earth's climate would change if the sun's output noticeably varied? Yet, the sun is unusually stable. Perhaps even more important, those stars probably vary with much larger sunspot activity than the sun has. Sunspot activity is accompanied by magnetic storms on the sun that produce harmful radiation. Presumably, stars that vary more than the sun does produce far more of this harmful radiation. Clearly, a stable sun is essential for life.

The Constellations

Astronomers divide the stars into groups that we call constellations. This word comes from the Latin words *con*, meaning "together," and *stella*, meaning "star." A constellation is a group of stars that are arranged in a sort of picture representing a person, animal, or object. There were 48 constellations handed down to us from the ancient Greeks, but since the Greeks borrowed heavily from earlier sources, many of those 48 constellations are far older than the ancient Greeks. No one knows exactly when or by whom those constellations originated, though a number of theories abound. Chapters 14–16 of this book examine one theory that is believed by many Christians. The ancient constellations encompassed the sky that was visible from about 35° north latitude a little more than 4,000 years ago. There were some gaps with few stars between those 48 constellations, and there is a large portion of the sky not visible from that latitude that had none of the classic 48 constellations. To fill those gaps and the large vacant area of the sky not visible from 35° north latitude, modern astronomers defined newer constellations, but most of the original 48 still survive.

The boundaries of the ancient constellations were not well defined, and there were competing constellations in the relatively empty parts of the sky. In 1930 astronomers officially fixed 88 constellations and defined constellation boundaries so that the entire celestial sphere is contained within those boundaries. Since the entire sky is filled with these 88 constellations, there will not be any more new ones. There are a few unofficial groupings of stars either within a single constellation or spanning several constellations. We call these asterisms. Examples of asterisms are the Big Dipper (part of the Big Bear) and the Summer Triangle (three bright stars in separate constellations). In addition to constellations, many stars have proper names. Though many star names appear to be of more recent origin than the original 48 constellations, there is some uncertainty about many of their origins and meanings. A few stars and constellations are mentioned in the Bible. We will discuss some of these here.

The Pleiades is neither a constellation nor an asterism. Instead, the Pleiades is a star cluster. The Pleiades appears as a tight knot of stars visible high overhead during winter evenings in the northern hemisphere. Six or seven Pleiads (what we call the individual members of the Pleiades) can easily be seen on a dark, clear night, though the star cluster contains about 1,000 stars. The Pleiades is sometimes called the Seven Sisters, and in Japan it is called

Subaru, from which the name of the auto company comes. The Pleiades are mentioned three times in the Bible, in Job 9:9 and 38:31, and in Amos 5:8. In Amos 5:8, the King James Version renders it "the seven stars," though more modern translations call it the Pleiades. The Hebrew word for the Pleiades that appears in all three passages is כִּימָה, which means a "heap" or "pile." The appearance of the Pleiades cluster is very distinctive, and nearly everyone has noticed them at some time or another.

In all three instances where the Pleiades is mentioned in the Old Testament, they are mentioned in conjunction with Orion. Orion is a large constellation consisting of many bright stars representing a hunter. Orion resembles the figure of a man, and it is relatively easy to pick it out. The Hebrew word used for Orion in the Old Testament is כְּסִיל (kᵉsîl), a word that is translated the other 70 times that it occurs as "fool." Notably, the word appears translated as "fool" 11 times in Proverbs 26. Jewish tradition identifies Orion with Nimrod, whom Genesis 10:8–9 calls a mighty hunter. However, record of this identification dates to medieval times, so it is not clear that the ancient Hebrews would have identified Orion with Nimrod. In all three of these Old Testament passages, credit is given to God as the one who made these groups of stars.

Job 38–41 is the LORD's answer to Job's protest, in which God presents Job with pointed questions. Within this response is Job 38:31, which reads,

> Can you bind the chains of the Pleiades
> or loose the cords of Orion?

Job himself had brought up the Pleiades and Orion in Job 9:9, so in this respect these words from God can be taken as a direct response to what Job had said much earlier. By asking this rhetorical question of Job, it is clear that only God is powerful enough to create the stars that we recognize as comprising the Pleiades and Orion.

But is there more to this? Some people have thought so. The Pleiades is a star cluster, which means that the stars are gravitationally bound. While the stars in a cluster orbit a common center of mass, most of the stars will remain close to one another for a very long time. While this is true of star clusters in general, the Pleiades is one of the closer star clusters to us, and it has a number of relatively bright stars. Hence, the Pleiades is the most easily recognized star cluster, and one of the few where individual member stars are so clearly visible. Due to their motions through space, stars gradually change position in the sky. Astronomers call this slow change in position proper motion. Extrapolating the proper motions of the Pleiads into the future shows that the Pleiades will remain a very recognizable pattern over the next million years or so. The same is true of the belt of Orion, another readily recognized pattern of stars. However, almost no other groupings of stars in the sky could be recognized in a million years from now. Therefore, some Christians have concluded that while man did not understand this until the 20th century, God knew it from the beginning, which is why God chose to use Orion and the Pleiades in Job 38:31. (A photograph of the Pleiades is on the front cover of this book, while a photograph of Orion is on the back cover.) However, there is a danger in positing such things, because Job 38:31 certainly had meaning to Job and the countless people who have read Job in the ensuing millennia. This is undermined if this true meaning was just recently revealed to man.

Furthermore, this approach has led to some embarrassing conclusions in the past. Johann Heinrich von Mädler (1794–1874) was a German astronomer who made a number of significant contributions, such as what was up to his time the most precise measurement of the length of the year. Unfortunately, von Mädler is best known for something else. In the mid-19th century, von Mädler determined that the Pleiades was the center of the universe, or more specifically, that Alcyone, the brightest star in the Pleiades, was the center of the universe. While this was dismissed within a few decades, how von Mädler reached his conclusion seemed sound enough at the time. Von Mädler built upon the work of William Herschel (1738–1822). By studying proper motions of stars, in 1783 Herschel determined that the solar apex, the direction in space where the sun is moving, was near the constellation Lyra.[2] A little later, Herschel determined with his "star gauging" that the sun was near the center of the Milky Way. At the time, many astronomers thought that the Milky Way was the extent of the universe, so the sun was at the center of the universe as well. In the early 20th century, both the ideas that the earth was the center of the Milky Way and that the Milky Way was the only galaxy would be overturned.

Von Mädler was a bit ahead of his time, for he differed with Herschel, believing instead that the sun was not at the center of the Milky Way, and hence must orbit the center. Von Mädler reckoned that the Pleiades coincided with center of the universe. While it is not clear how he came to this conclusion, we may surmise. Because the sun orbits close to the galactic plane, we would expect that the galactic center ought to lie close to the plane of the Milky

[2] The modern determined position of the solar apex is displaced a bit from this position.

Way. Furthermore, the center of the galaxy must lie near a circle making a 90° angle to both the solar apex and solar antapex. There are two places where this circle and galactic plane intersect. The Pleiades lies along the circle near one of the intersections with the galactic plane. Von Mädler selected the incorrect intersection, for the modern determination of the galactic center is near the other intersection, about 180° from the Pleiades.

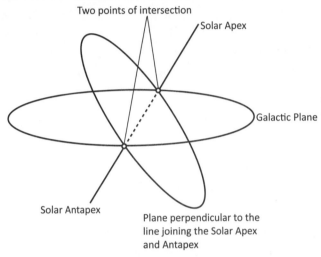

A number of Christians and other writers picked up on Von Mädler's view, and interpreted it in terms of Job 38:31. For instance, Frances Rolleston, who created the "gospel in the stars" theory (discussed in Chapters 14–16) argued that the name Alcyone means "the center,"[3] and thus supposedly proved that primeval man had cosmological knowledge that had been lost only to be rediscovered in more modern times. Other supporters of the gospel in the stars continued to repeat this nonsense even after astronomers had repudiated the notion that Alcyone was the center of the universe.

Another victim of von Mädler's error was the early Mormon apologist George Reynolds (1842–1909). In 1879, Reynolds published *The Book of Abraham: Its Authenticity Established as a Divine and Ancient Record: With Copious References to Ancient and Modern Authorities.* As the name suggests, in this work Reynolds defended the authority of *The Book of Abraham*, a text

---

[3] It does not mean this. Rather, in Greek mythology Alcyone was the name of one of the Seven Sisters, the daughters of Atlas and Pleione. The association of Alcyone and the other Seven Sisters with individual stars in the Pleiades star cluster came during the Renaissance, not ancient times.

sacred to Mormons supposedly translated by Joseph Smith from Egyptian papyri in 1835. Reynolds included a brief description of Alcyone's (literal) central importance. Reynolds also discussed the relationship of Alcyone to the Great Pyramid of Giza. Pyramidology was very popular at the time, and as a result many people became adherents. For instance, in 1877 Joseph Seiss, an early popularizer of the gospel in the stars, wrote *A Miracle in Stone, or The Great Pyramid of Egypt*. While pyramidology is not as popular as it once was, fascination with this largely bogus belief persists today. And Orion frequently plays a prominent role in pyramidology even today. The modern incarnation of this is called the Orion Correlation theory, and is attributed to Robert Bauval (b. 1948). However, the connection of Orion to pyramidology goes back a century and a half, and probably Orion was tied to the pyramids via its connection to the Pleiades in Job 9:9 and Job 38:31.

Another oddity related to Orion is the belief among some Seventh-day Adventists that God's throne is situated in or near the sword of the constellation Orion. This belief stems from a December 16, 1848, vision that the Adventist "prophetess" Ellen G. White (1827–1915) had. Of her vision, White wrote,

> Dark, heavy clouds came up and clashed against each other. The atmosphere parted and rolled back; then we could look up through the open space in Orion, whence came the voice of God. The Holy City will come down through that open space.[4]

To some, the mention of an open space is similar to the empty place of Job 26:7, which states that God stretched out the north over an empty place and hung the earth upon nothing. Even some non-Seventh-day Adventist preachers and teachers have become excited about this. One problem with equating what White and Job stated is that Orion, lying slightly below the celestial equator, could hardly be considered to be in the north. The Holy City here is the New Jerusalem as described in Revelation 21. Of particular interest is Revelation 21:10, which says that the city will descend out of heaven from God. Apparently, this is what White was referring to, as many other Seventh-day Adventists understand this to be the meaning.

---

[4] This quote is contained in an entry in White's writings entitled "To Those Who Are Receiving the Seal of the Living God," penned January 31, 1849 (http://text.egwwritings.org/publication.php?pubtype=Periodical&bookCode=Broadside2&lang=en&collection=2&section=all&year=1849&month=January&day=31).

The source of this idea almost certainly is a pamphlet published by Joseph Bates (1792–1872) in 1846, two years prior to White's vision. Bates was an early important figure in Seventh-day Adventism, and he influenced James and Ellen White, but they also influenced him. Bates' pamphlet, entitled "The Opening Heavens," amounts to a defense of the literal nature of the yet future descending New Jerusalem. Many Christians today view this as literal, but in Bates' time many viewed the descending New Jerusalem figuratively. Bates was a survivor of the Millerite Great Disappointment two years earlier in 1844; but undeterred, he argued for the arrival of the New Jerusalem as imminent. Bates mixed in with his treatise astronomers' descriptions of the Orion Nebula as then known. Some of those descriptions, based upon the view through a telescope, were of an opening in the black of the sky beyond which light appeared to shine. Armed with these descriptions, it was easy to view the glow of the Orion Nebula as the radiance of God's glory. Bates also noted that the Tree of Life is mentioned only twice in the Bible, in the Garden of Eden in Genesis 2–3 and again in the New Jerusalem in Revelation 22. Finally, Bates repeatedly pointed out that after the Fall, God prevented Adam's access back to the garden and the Tree of Life with cherubim and a flaming sword. Bates' pamphlet concludes,

> This then is the capacious and glorious "golden City;" the "New Jerusalem;" the "heavenly Sanctuary;" the "Bride of the Lamb's Wife;" the "Mother of us all;" the "Paradise of God;" the capital of our coming Lord's EVERLASTING kingdom, which is now about to descend from the "third heaven" by the way of the open door, down by the "flaming sword" of Orion. O let us see to it, that we are all ready to enter into this celestial City.

That is, an important part of Bates' identification of the Orion Nebula as the direction from which the New Jerusalem will come is the fact that the way to the Tree of Life was guarded by a flaming sword, and the Orion Nebula has a fiery appearance, and is located in the sword of Orion! Since the New Jerusalem will come from God, one could easily infer that the direction in space from which the city comes must be the direction of God's throne. Ergo, the Orion Nebula is the direction to God's throne. Today, this reasoning may seem silly to many people, but given what was known about the Orion Nebula then and the theological environment in which Bates and early Seventh-day Adventists were, this conclusion does not seem so absurd. However, this is not something that Seventh-day Adventists frequently talk about today.

In addition to the Pleiades and Orion, Job 9:9 and Job 38:32 (which follows Job 38:31) also mention Arcturus (as translated in the King James Version). Arcturus is a bright star in the constellation Boötes. However, the English Standard Version, New International Version, and New American Standard Bible render this as "the Bear," referring to the constellation of the Big Bear (Ursa Major). The Hebrew word used here is עַיִשׁ (*'ayiš*). Nearly all commentators think that the intended association here is with the Big Bear and not with the star Arcturus. Arcturus is near the Big Bear. In fact, the easiest way to find Arcturus in the sky is to follow the arc in the handle of the Big Dipper (or the tail of the bear) away from the Dipper to Arcturus (we say, "arc to Arcturus"). Because of its close proximity to the Big Bear and the Little Bear (Ursa Major and Ursa Minor), Arcturus has long been known as the "keeper of the bears," from which its name likely arises. While *ursa* is the Latin word for "bear," the Greek word for "bear" is *arctos*. Our word *arctic* derives from an ancient belief that bears had something to do with the far north. Job 38:32 appends to the mention of Arcturus "and her sons," or decedents. This could refer to the constellation of the little bear or stars associated with the big bear, or to both bear constellations if the star Arcturus is the intended meaning, though this is conjecture. We ought to mention that a few commentators even think that *'ayiš* actually may refer to the Hyades, another star cluster near the Pleiades. So there is some uncertainty as to exactly what *'ayiš* refers to here. The exact meaning is not as important as understanding that, as with the Pleiades and Orion, the passages in Job are giving honor to God as the Creator of all that we see.

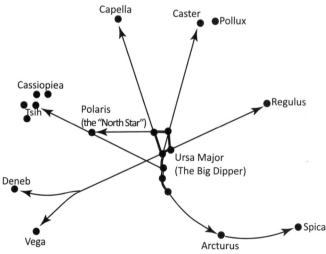

Job 9:9 also mentions "the chambers of the south." The meaning is uncertain, but it likely refers to a group of constellations seen along the southern horizon at the time of Job.[5] Another astronomical term found in Job 38:32 is "mazzaroth." Mazzaroth is a transcription of the plural Hebrew noun מַזָּרוֹת (*mazzārôt*), which appears only once in the Old Testament. The etymology of *mazzārôt* is unclear and scholars are not sure exactly what *mazzārôt* means.[6] It likely refers either to constellations in general or to the zodiacal constellations. It may also refer to a particular star, but this is unlikely, in view of the use of the plural.

A similar term, מַזָּלוֹת (*mazzālôt*), is found in 2 Kings 23:5, but appears merely to be an alternate spelling of *mazzārôt*. The King James Version translates *mazzālôt* in 2 Kings 23:5 as "planets," though this is almost certainly incorrect, or at least too restrictive. The English Standard Version, New American Standard Bible, and New International Version translate it as "constellations," as most commentators agree that *mazzālôt* refers either to the constellations in general or to the zodiacal constellations. Often missed is the fact that Job 38:31–32 is followed by discussion of weather and seasonal changes. This is a common theme found in ancient treatments of stars and constellations. Even today farmers' almanacs include descriptions of astronomical phenomenon and weather forecasts. These stem from a long-understood knowledge that the changing weather and seasons were accompanied by changes in the constellations visible throughout the year. This is in keeping with the God-ordained function of astronomical bodies for the reckoning of time and seasons. Job clearly gives God credit for all of these changes, but unfortunately, many ancient cultures took their eyes off the Creator and began to think that the astronomical bodies brought about seasonal weather changes. The belief persists today in a modern day form of astrology. (Chapter 5 discusses astrology in more detail.)

The phrase "host of heaven" appears in 2 Kings 23:5 and refers to stars as a whole, as it does in 13 other Old Testament passages and in one New Testament passage. Examples include Deuteronomy 4:19 and Acts 7:42 to refer to stars as objects of worship by pagans, and Isaiah 34:4 and Jeremiah 33:22 as a description of stars. Three other times "host of heaven" refers to angels, as in

---

[5] An effect called precession now renders some of those constellations continually below the southern horizon from the likely latitude of Job.

[6] Ignorance of the meaning probably explains why it was transliterated rather than translated.

Nehemiah 9:6. The context of Acts 7:42 mentioned above is Stephen's speech that led to his martyrdom. Acts 7:42–43 is a quote from Amos 5:25–27. Amos 5:25–26 reads,

> Did you bring to me sacrifices and offerings during the forty years in the wilderness, O house of Israel? You shall take up Sikkuth (סִכּוּת) your king (מֶלֶךְ), and Kiyyun (כִּיּוּן) your star-god (כּוֹכַב אֱלֹהֵיכֶם)—your images that you made for yourselves.

In Acts 7:42–43, Stephen quoted from the Septuagint, which reads סִכּוּת as "tabernacle" and מֶלֶךְ as "Molech." Furthermore, the Septuagint interpreted "Kiyyun" (כִּיּוּן) as "Remphan." Who or what was Kiyyun/Remphan? There is some uncertainty, but it probably refers to an idol that was the embodiment of a deity represented by the planet Saturn. Kiyyun probably derives from the Babylonian name for Saturn, and Remphan is the Greek equivalent. Likewise, Sikkuth is similar to the Akkadian name for Saturn. However, without vowels, it is also similar to the (plural) Hebrew word for tabernacle(s), which probably is why the Septuagint and King James Version translated it as "tabernacle." With recognition of the likely origin of these words, the passage from Amos 5:26, "take up Sikkuth your king, and Kiyyun your star-god" amounts to a parallel statement, noting the worship of Saturn, with Saturn referred to by both the Babylonian and Akkadian names.

In Isaiah 13:10 the plural form of the Hebrew word $k^e s \hat{\imath} l$ is used in reference to the constellations. This is the Hebrew word for Orion. Orion is a prominent constellation, so here it is a sort of stand-in for constellations in general. We do a similar thing in English. For some time after the launch of *Sputnik*, the first artificial satellite in 1957, new satellites that were launched were called "sputniks."

The "crooked serpent" of Job 26:13 likely refers to a constellation.[7] The first part of the verse states that God's spirit, or breath, garnished the heavens; the second part says that God's hand formed the crooked serpent. Some identify this serpent with the constellation Draco. Today we generally view Draco as a dragon, but the ancient languages here can be confusing to our modern uses. Dragons, serpents, and other reptile-like creatures were frequently grouped together. There are other serpents or serpent-like creatures amongst

---

[7] "Crooked serpent" is the translation of the Kings James Version; the English Standard Version and New American Standard Bible have "fleeing" instead of "crooked" in both Job 26:13 and Isaiah 27:1.

the constellations, so this passage may refer to one of those. Isaiah 27:1 also mentions a crooked serpent, but this context does not suggest a constellation.

Stars are mentioned in a number of different contexts in the Bible. The first mention is of course at their creation in Genesis 1:16. The second mention is in Genesis 15:5 when God established His covenant with Abraham and told Abraham that the stars are innumerable. In the second century, Claudius Ptolemy catalogued 1,022 stars. During the Middle Ages there arose a belief that the ancient authorities such as Aristotle and Ptolemy knew everything, and since Ptolemy catalogued 1,022 stars, that was all the stars that there were. Of course, Ptolemy never made such a claim, and this conclusion is contrary to what the Bible actually teaches as well. Psalm 147:4 tells us that God counts the stars and calls them all by name. How can one count what is innumerable? The easy answer is that all things are possible with God. However, being innumerable does not mean that there are an infinite number of stars. Instead, it means that there are so many stars that it is impossible for man to count them, but God in his infinite wisdom can count them.

Stars are mentioned again in Genesis 37:9, in one of Joseph's dreams. Numbers 24:17 contains Balaam's famous messianic prophecy about a star coming forth from Jacob. In the song of Deborah and Barak, Judges 5:20 records that the stars fought against Sisera. It is obvious that the stars in these three passages are not literally stars. This non-literal understanding likewise appears in Job 38:7 where it states that the morning stars sang at creation. The morning stars most likely are angels. Probably on the basis of this verse and Revelation 12:4, some Christians have suggested that stars are not physical entities at all but rather are the manifestations of angels. But then Daniel 12:3 likens the saints to stars, so to be consistent one would have to argue that stars are saints as well. It is thus best to accept these as figurative uses. On the other hand, some sort of literal star led the magi to the young Jesus (Matt. 2:7–10). In defending the doctrine of the resurrection of the dead, the Apostle Paul wrote in 1 Corinthians 15:41 that glory of the stars differs from the glory of the sun and the moon and that even stars differ in glory. We have already seen that Jude 13 compares apostates to wandering stars and that this is the exact term that Greeks applied to planets. Some have suggested that Jude's warning may compare apostates to meteors or comets, but this is unlikely.

The book of Revelation uses stars in a symbolic sense to refer to the messengers of the seven churches (Rev. 1:16–20; 2:1; 3:1). It also refers to Jesus as "the bright and morning star" (Rev. 22:16). The 12 stars in the crown of the woman in Revelation 12:1 probably represent the tribes of Israel. This association is anticipated by Joseph's dream in Genesis 37, where 11 stars represent Jacob's other sons, from which, along with Joseph, each of the tribes of Israel descend. Furthermore, verse 2 reveals that she is in the process of giving birth. There are several Old Testament passages that allude to Israel (or Judah) as a woman giving birth (Isa. 26:17–18; 66:7–9; Mic. 4:10; 5:3). Most commentators think that the third of the stars that Revelation 12:4 states were swept away and thrown to earth means angels that followed Satan in his rebellion. Elsewhere Revelation appears to mention stars in a literal sense.

A new view about Revelation 12:1–2 has come about in recent years. While agreed that this refers to the birth of Jesus, this new view argues that these two verses reveal the time of Jesus' birth. This theory identifies the woman clothed with the sun as the constellation Virgo, with Virgo representing the Virgin Mary. The crown of 12 stars does not represent Israel, but rather is literally 12 stars near the head of the constellation Virgo. The moon beneath her feet literally is the moon. For that matter, being clothed with the sun refers literally to the sun's location in the constellation Virgo. The sun is in the constellation Virgo for approximately a month each year. In the 21st century, the sun is in Virgo from mid-September to mid-October, but 2,000 years ago it would have been mid-August to mid-September. Hence, if this theory is correct, it would place the birth of Jesus later in summer, shortly before the autumnal equinox. The moon is near the feet of the constellation Virgo one day per month, so the circumstances assumed here will happen one day per year, again in late August or early September. Therefore, this theory purports to fix the date of Jesus' birth, assuming that one knows the year. This view about Revelation 12:1–2 has become common among those arguing for the late death of Herod, allowing for Jesus to be born in 3–2 BC (see Chapter 8). Within this theory, an oft-cited date of Jesus' birth is September 11, 3 BC. If Passover were in mid-March that year, then this could have been the date of *Rosh Hashanah*, the Jewish New Year. Since the shofar is blown on this day, this is the Feast of Trumpets. By Jewish tradition, *Rosh Hashanah* is the anniversary of the creation of the world. Because the Crucifixion was at Passover and the coming of the Holy Spirit was at Pentecost, some people think it fitting that the Messiah was born at *Rosh Hashanah*. Incidentally, many people who support this theory also think it likely that Jesus will return on *Rosh Hashanah*.

How plausible is this theory? It has several obstacles to overcome. First, one is hard-pressed to find any commentaries that support this position. This is because this is such a new idea. If one sees a new meaning in a biblical passage that no one has seen before, it likely is because it is not there. Therefore, one ought to be cautious about this theory. Second, this alleged sign largely would not have been visible. The stars of Virgo would not have been visible, because the sun was there. The moon, if visible, would have been barely visible, and then only briefly as a thin crescent low in the sky shortly after sunset. There would have been no stars visible with it. Since this "great sign" (the Apostle John's term) was not readily visible, this great sign is at best inferred by interpretation. Note that interpretation here refers to identifying the sign, not revealing its meaning. Third, we have no ancient records of Jewish astronomical knowledge or lore. Medieval Jewish literature identifies the constellation Virgo as a virgin, but this came very late, so it is not at all clear the Apostle John or any other Jews in the first century would have made this connection. Hence, this supposed sign appears not to have gained any attention until two millennia after it happened. Fourth, identification with the particular date of September 11, 3 BC for Jesus' birth requires further assumptions. It requires that the month of *Aviv* began relatively early that year, about a week before the vernal equinox. As will be discussed in Chapter 4, we do not know what methodology for intercalation of a month that the Jews used 2,000 years ago, so exactly when the ceremonial year began some years, such as this one, is uncertain. If the month of *Aviv* began in April that year, then *Rosh Hashanah* would have been in October. Furthermore, if the Jews observationally determined the beginning of each month, then the thin crescent moon on September 11, 3 BC may not have been visible, thus placing this date one day off. Given these considerations, Christians are advised to tread lightly with this theory.

The stars of heaven falling to the earth in Revelation 6:12 occurs with the breaking of the sixth seal, and is accompanied by great tumult in the physical world. This upheaval includes a great earthquake, the sun turning black, the moon becoming as blood, the sky rolling up like a scroll, and the mountains and islands moving out of place. The falling of the stars here most likely refers to a great meteor shower. In today's modern scientific definitions we don't view meteors as stars, but we still commonly refer to meteors as shooting stars or falling stars. At the sounding of the fourth trumpet, Revelation 8:12 records that a third of the sun, the moon, and the stars are struck so that a third of their light was darkened. Apparently, the light of the sun and the moon are reduced by one-third, and so it follows that the stars will dim by as much. We don't know whether this is a physical change in the astronomical bodies themselves or if the earth's atmosphere will be altered. Either way, this means a reduction in the light of real stars.

The Pale Blue Dot

The Bible gives several purposes for the heavenly bodies. However, in a secular worldview, these purposes are entirely missed. In 1994 Carl Sagan published a book entitled *The Pale Blue Dot*. The title was inspired by a 1990 photograph made by Voyager I after it had completed its mission of studying the outer planets and was more than six billion kilometers from earth. Controllers turned the spacecraft back toward the solar system to photograph all of the planets.[8] From the distant vantage of Voyager I, the planets appeared relatively close to the sun. From this distance, the earth spanned less than one pixel in its photograph, and because of the earth's close proximity to the sun, reflections of the sun's light appeared as beams crossing the photograph. The earth's image was a tiny blue dot intersected by one of those beams. This photograph soon became known as the pale blue dot, and it triggered thoughts about how insignificant we and our world are in the vast cosmos. All of our cares and our concerns are found on this speck that is tiny when compared to the rest of the universe. We think that we are important, but the scale of the universe suggests otherwise.

In his book, Sagan asked what our significance was and attempted to attach some meaning to our existence. The problem is, Sagan's worldview was that of agnosticism. But without God, our lives have no significance and no purpose. Neither does the universe. Ultimately, all is meaningless. To attempt to find purpose where there is none is a fool's errand. Most people find the supposed meaninglessness of our existence very unsettling. Sagan obviously wrestled with this. Ultimately, he concluded that mankind has accomplished some worthwhile things (though it is not clear what that even means), and it would be a shame if we were to snuff that out by destroying ourselves.

Three thousand years earlier the psalmist David raised these same questions. In Psalm 8:3–4 David wrote,

---

[8] Pluto, which still was considered a planet in 1990, was omitted, because it was far too faint to photograph. In 2006, astronomers defined a planet for the first time, which effectively excluded Pluto from being a planet.

When I look at the your heavens, the work of your fingers,
  the moon and the stars, which you have set in place,
what is man that you are mindful of him,
  and the son of man that you care for him?

David understood that we are surrounded by a very large and marvelous creation, and that when compared to that creation, we are pretty insignificant. However, instead of relying upon his own wisdom to develop some importance for humanity, David went on to note in the next four verses:

Yet you have made him a little lower than the heavenly beings
  and crowned him with glory and honor.
You have given him dominion over the works of your hands;
  you have put all things under his feet,
all sheep and oxen,
  and also the beasts of the field,
the birds of the heavens, and the fish of the sea,
  whatever passes along the paths of the sea.

By recognizing that we have a Creator, David reached a very different conclusion from Sagan. We do have purpose. As the Westminster Catechism states, the chief end of man is to glorify God and enjoy Him forever. The whole of the Bible unfolds God's plan of salvation for man. Only through our redemption through the finished work of Jesus Christ and our submission to God's will can we find the true intended purpose for our lives.

CHAPTER 4

Biblical Teachings about the Purpose of the Heavenly Bodies

The Day Four creation account (Gen. 1:14–19) gives several purposes for the heavenly bodies. One of the purposes given in Genesis 1:14 is a basis for marking time and hence developing calendars ("let them be for . . . seasons, and for days and years"). The days and years may be plain enough, but many people get the wrong impression with the mention of seasons. Most people think of seasons in terms of the climatic seasons: spring, summer, autumn, and winter. However, the word *season* has other meanings as well. One possible meaning is a period of time appointed for some purpose. For example, we refer to seasons for sports and other activities in this way. There is baseball season, football season, and basketball season. Furthermore, there are seasons for hunting various game animals. This is the sense of the word *season* in Ecclesiastes 3:1, which says,

> For everything there is a season, and a time for everything under heaven.

Admittedly, while the word *season* appears in English translations of both Genesis 1:14 and Ecclesiastes 3:1, different Hebrew words are used in these two verses (מוֹעֵד versus זְמָן). However, the semantic ranges of these two Hebrew words overlap, and the English translation as "season" in either case is justified. Elsewhere in the Old Testament, the Hebrew word translated "seasons" in Genesis 1:14 is used to refer to the various festivals that the Hebrews were to observe, such as Passover (cf. e.g., Exod. 13:10). These festivals were observed on particular days of particular months, so Genesis 1:14 not only refers to days and years, but to months as well. This is reinforced by Psalm 104:19, which states,

He made the moon to mark the seasons;
　the sun knows its time for setting.

Leviticus 23 details the timing of each of the Hebrew feasts. The feasts occur on specified days of particular months. Since the moon defines the beginning and duration of the months, the moon plays an integral role in the seasons mentioned in Genesis 1:14.

The Natural Units of Time

The day, month, and year are natural units of time measurement. They are natural in that they are defined in terms of the motions of heavenly bodies. The day is the rotation period of the earth. The earth's rotation causes the sun to appear to move across the sky during the course of the day. The year is the revolution period of the earth. Due to the tilt of the earth's axis, our climatic seasons repeat each year, so the year is an important basis for agriculture. In between the day and year is the month, the orbital period of the moon around the earth. Other divisions of time, the week, the hour, the minute, and the second are not natural units, because they do not have their basis in the movements of the astronomical bodies.

A complication arises in that the day is not divisible into the month or year, nor is the year divisible by the month. There are various ways of handling this difficulty, which results in different calendars. There probably is no single, God-ordained calendar. In this chapter, we will explore the subtleties of time measurement. Along the way, we will see what sort of calendar that we think the ancient Hebrews used. We will trace the history of the calendar that is in general use today, and we will see why Jewish festivals appear to bounce around on the modern calendar.

The year is the revolution period of the earth around the sun. Motions, such as the earth's revolution around the sun, must be measured with respect to some reference. Different reference points will result in different definitions of the year. The *sidereal year* is the orbital period of the earth with respect to the stars (365.25636 days).[1] That is, if the earth, the sun, and a distant star align with one another, we say that one sidereal year has elapsed when the three align once again. The sidereal year is the true orbital period of the earth, because the stars are so distant as to represent a good nonmoving standard of measurement. This is the year that physicists and astronomers must use as the basis of physical computations when they want the most precise results.

---

[1] The word *sidereal* means "star."

The *tropical year* is the revolution period of the earth with respect to the vernal equinox (365.24219 days). The equinoxes are the intersections of the ecliptic (the earth's orbital plane) and the celestial equator (an imaginary circle in the sky lying directly above the earth's equator). The ecliptic and celestial equator intersect, making an angle of 23.4°, the angle of the earth's axial tilt. Because both the ecliptic and celestial equator are great circle arcs, they intersect in two places, and hence there are two equinoxes. The equinox where the sun crosses the celestial equator traveling northward is the vernal equinox; the equinox where the sun crosses moving southward is the autumnal equinox.

Why are the sidereal and tropical years not the same length? The spinning earth has a slight equatorial bulge. The gravity of the sun and other objects in the solar system produces a torque on this bulge. This torque results in a gradual shift in the orientation of the earth's rotation axis, an effect that we call *precession*. The earth's axis gradually sweeps out a cone. Precession is easy to demonstrate with a spinning top or gyroscope. The ecliptic is reasonably fixed, but the celestial equator is perpendicular to the rotation axis of the earth and hence must precess as the rotation axis precesses. Thus, the intersections of the celestial equator and the ecliptic, the equinoxes, gradually shift along the ecliptic. In fact, astronomers call this the precession of the equinoxes. It takes 25,900 years to complete one precession cycle.

Since the vernal equinox slides very gradually along the ecliptic or against the background stars, the tropical year and the sidereal year cannot have the same length. Which year is the basis of our calendar? The sidereal year is the true orbital period of the earth, but the seasons repeat with the tropical year. Until recently, virtually all societies were agrarian and hence were directly dependent upon agriculture. Even today, we must eat, so we are still dependent upon agriculture, albeit less directly. Knowing when to plant crops is essential knowledge in farming, and so the tropical year is the basis of our calendar. This keeps the seasons synchronized to our calendar, which even most modern people find agreeable.

We can mention at least one other year, the *anomalistic year* (365.25964 days). The anomalistic year is the revolution period of the earth with respect to perihelion. Perihelion is the point on a planet's orbit that is closest to the sun. The gravitational perturbations of the other planets cause the earth's perihelion to gradually shift along the ecliptic as well, an effect we call *perihelion precession*. For most purposes, the anomalistic year is not nearly as important as the other two years.

The month is the orbital period of the moon, but as with the year, we must define the month with respect to some reference. The *sidereal month* is the orbital period of the moon with respect to the stars. Since the stars represent a distant, reasonably fixed reference, the sidereal month is the true orbital period of the moon. However, the *synodic month* is the more obvious orbital period of the moon. The synodic month is the orbital period of the moon with respect to the sun. Since the geometrical relationship between the moon, the sun, and the earth determines lunar phases, the synodic month is the period in which the lunar phases go through a complete cycle.

The sidereal month is 27⅓ days, while the synodic month is 29½ days. Why are the sidereal and synodic months different? During the course of a month, the earth moves nearly one-twelfth of its orbit around the sun. It takes the moon a little over two days to make up this difference in distance. The moon's changing phases are very obvious, much more obvious than the moon's changing position with respect to the stars, so the synodic month is the basis of the month on all calendars.

There are other ways of defining the month as well, such as the *nodal month* (27.21222 days). The nodal month is the orbital period of the moon with respect to its nodes, the nodes being the intersection of the moon's orbit and the ecliptic. The nodal month is important in predicting eclipses, but it is the synodic month that is normally the month of choice for calendar purposes.

The day is defined as the rotation period of the earth, but as with the year and month, we must specify the reference frame. The most obvious references are the sun and the stars. The *solar day* is the rotation period of the earth with respect to the sun, and that is the day that we normally use. The *sidereal day*, the rotation period of the earth with respect to the stars, is the true rotation period of the earth. The sidereal day is about four minutes shorter than the solar day. The extra four minutes is made up by the motion of the earth around the sun over the course of a day. The result is that stars rise about four minutes earlier each (solar) day.

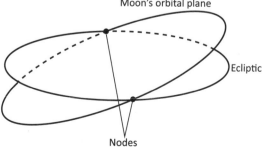

Calendars

Most ancient calendars were observationally based. All such ancient calendars considered the beginning of the month to be at the new moon. At new moon, the moon is not visible for two to three days. Thus there were different ways of defining a new moon. In some systems, the morning of last visibility of the waning crescent moon or the first morning of invisibility of the waxing crescent moon served to indicate a new moon. However, a more common method of defining new moon was the first visibility of the waxing crescent in the evening.

For example, on the Hebrew calendar, which is still observed today, a month begins when one can spot the first thin crescent moon in the western sky following a new moon. Since this is done shortly after sunset, and the Hebrews reckoned the beginning of the day from sunset, this observation quickly determined whether a particular day that just started was the first of a new month or the final day of the previous month. The moon generally is not visible for two to three days near new moon. Since the synodic month is approximately 29½ days in length, if the beginning of each month (and likewise the end of the preceding month) is observationally determined in this way, the months generally will alternate between 29 and 30 days. With a few years of records, it is possible to anticipate when the first thin crescent is likely to be visible, and thus one can calculate with some certainty in advance when a month is likely to begin (or when a month in the past would have begun).

The tropical year is not divisible by the lunar month, so a strictly lunar calendar rapidly gets out of step with the year. Twelve lunar months will be about ten days shorter than the tropical year. For example, the Islamic calendar is a lunar calendar, so the months of that calendar slip ten days earlier each year. This is why the Islamic holy month of Ramadan occurs progressively earlier each year, so that month of religious observance is not fixed with respect to the climatic seasons. The simplest fix for this problem is to adopt a lunisolar calendar. Because the discrepancy is ten days each year, after three years the accumulated difference from the tropical year is 30 days, or about one month. Therefore, insertion of an extra month, called an intercalary month, approximately every third year brings the calendar back in step with the tropical year. Usually, the intercalary month was placed at the end of the year. For many ancient calendars, the end/beginning of the year was near the time of the equinoxes. A year with an intercalary month was approximately 385 days long, and years without an intercalary month were approximately 355 days long.

Inserting an intercalary month every third year, however, does not fully compensate for 12 synodic months falling short of the tropical year. A much better match is to insert an intercalary month seven times in 19 years (as opposed to six times in 18 years). We call this 19-year pattern the Metonic Cycle, because Meton of Athens, a 5th century BC mathematician and astronomer, discovered and wrote about it. (Actually, cultures around the world independently discovered the Metonic cycle, and the first discovery *could* have predated Meton.) The Metonic cycle inserts an intercalary month in years three, six, eight, 11, 14, 17, and 19 of a 19-year cycle. Many ancient calendars followed this method. The Jewish calendar observed today also uses the Metonic cycle. We do not know when the Jews adopted this approach, but many think that it may have happened during the Babylonian captivity. There are no direct biblical references to how intercalation was done in Old Testament times, and the earliest specific mentions from secular Jewish literature are medieval.

The Romans took a different approach. The decision of when to insert an intercalary month was left to priests. However, the priests often succumbed to political pressures in their intercalation decisions. The terms of many political appointments were expressed in years, so it was advantageous to withhold adding intercalary months when political opponents were in office, effectively shortening their terms. Therefore, intercalation often did not happen for many years in succession. On the other hand, it was politically expedient to make up those extra months when political allies were in office, so occasionally years had more than 13 months. Adding to the confusion, many of these decisions were local, so what month and year it was depended upon where one was in the Roman Empire. To rein in the confusion, Julius Caesar instituted a calendar reform in 45 BC.

The most significant change of the Julian calendar reform was to scrap the lunisolar calendar in favor of a solar calendar. This required increasing the length of the year from 355 days to the more correct length of 365 days. This was done by distributing an extra ten days to the 12 months that then existed in the calendar. This removed the need for an intercalary month. This also meant that the phases of the moon now drifted progressively earlier in each succeeding month, thus abandoning the strictly lunar basis of the month.

A second very important change of the Julian calendar reform was the institution of observing leap days. Since the tropical year is approximately 365¼ days, an extra day inserted at the traditional place of the intercalary month brought the calendar year into near alignment with the tropical year. The intercalary month had formerly been placed between February and March, so this was where a leap day was inserted every fourth year, and eventually people came to include this extra day as part of February.

Most readers will recognize the elements of this calendar in our own calendar. Some readers probably are aware that the tropical year is actually less than 365.25 days, which results in an error of ¾ of a day per century. The Council of Nicaea in AD 325 is famous for its creed that clearly defined orthodox Christianity as opposed to many heresies that had arisen by that time. Less well known is that the Council of Nicaea also established when certain Christian holidays, such as Resurrection Sunday, were to be observed. Due to the slight mismatch between the Julian calendar and the year's actual length, by the 16th century there was an accumulated error of ten days. Not only were the seasons creeping earlier in the calendar, but the observance of Resurrection Sunday, as determined by the formula adopted by the Council of Nicaea, was too. This situation was unsatisfactory, so Pope Gregory XIII commissioned a reform of the calendar in 1582. The Gregorian calendar essentially is the calendar that we use today.

What did the Gregorian calendar reform do? First, to bring the calendar back to the standards of AD 325, the Gregorian calendar reform deleted ten days from the calendar—October 4, 1582, was immediately followed by October 15. The more lasting legacy of the Gregorian calendar reform was that it altered the rule for adding leap days. Under the Julian calendar, any year divisible by four is a leap year, but even century years (divisible by 100) are not leap years (they are common years), unless they are also divisible by 400. This omits three days per four centuries, which amounts to the ¾ day discrepancy for the Julian calendar. Thus, while the years 1700, 1800, and 1900 were not leap years, 1600 and 2000 were.

A less well-known legacy of the Gregorian calendar reform is that it moved New Year's Day from the time of the vernal equinox to January 1. The Gregorian calendar reform took effect during the Protestant Reformation, so many non-Roman Catholic countries did not adopt the new calendar until years later. This caused much confusion. Adding to that confusion, even where the Gregorian calendar was adopted, some people continued to observe New Year near the vernal equinox. In some countries, New Year had been April 1. Derision for those using the old calendar is the origin of the term "April fools."

We should emphasize that there are many other ways that one could reconcile what appears to be a mismatch between the lengths of the days, months, and years; and different ancient societies used different methods. It is important to realize that there is not a single, uniquely satisfying way to do this, or else there would not be such diversity. As with any other measurement system, we get very comfortable with what we are used to, and think other measurements are odd or downright weird.

For instance, many moderns find the Hebrew calendar odd with the dates of Passover, *Yom Kippur,* and other festival and holy days moving about our calendar. However, on the Jewish calendar, those observances are on the same dates each year. As an example, Passover is on the 15th day of the first month of the religious calendar. The 15th day of any strictly lunar month will be the full moon, and the first month is the month that follows the vernal equinox. The Jewish New Year (on the civil calendar) is the first day of the seventh month of the religious calendar, which occurs about the time of the autumnal equinox. Jewish tradition holds that the creation was at this time, so this is a logical choice for the beginning of the year. The religious calendar was introduced at Sinai as a memorial to the first Passover. The Jewish civil calendar almost certainly predates the religious calendar. Anyone who is used to a lunar calendar would find it odd that our celebrations do not occur on the same phase of the moon during the respective month each year. In other words, our fixed dates bounce about on a lunar calendar.

The One God-Ordained Unit of Time: The Week

While the day, month, and year are natural units of time, there is no indication in Scripture that God directly ordained those units of time be observed. One could argue that since God created the astronomical bodies and, since God certainly intended their use as time markers in Genesis 1:14, then there is at least an implicit ordination of the day, month, and year. However, unlike the day, month, and year, there is no natural basis for the week. The week clearly is divinely ordained in Scripture. Genesis 1 records that God created the world over six days. Genesis 2:1–3 states that God rested on the seventh day, and that He blessed the seventh day and made it holy. This is reiterated in the giving of the Fourth Commandment in Exodus 20:8–11 (cf. Deut. 5:12–14), as well as in Leviticus 23:3. Thus, the observance of the week is God's plan and command.

Most historians think that the seven-day week originated in Mesopotamia, going back at least as far as the Akkadian Empire. The Egyptians supposedly adopted the seven-day week from Mesopotamia. The Romans in turn picked up the practice of the seven-day week from the Egyptians early in the Christian era, abandoning the eight-day week that they previously used. According to this secular view of history, the ancient Hebrews came to this rather late too, having learned of the seven-day week either from Mesopotamia or Egypt. However, according to the Bible, the pattern of a seven-day week goes back to Genesis 1. The establishment of the Sabbath as such came with the giving of the Law in the 15th century BC. The sense of Exodus is not that God established a new system for the observance of a weekly day of rest. Rather, it comes across as God explaining the significance of why it was to be a day of rest, and an explanation of why the Sabbath was to be kept holy. The absence of direct ordination of a new system for the workweek suggests that Israel already had kept a seven-day week (possibly with a Sabbath) in Egypt. At the same time, a seven-day week could have originated in Mesopotamia as a result of transmission of oral tradition passed down through the pre-Flood patriarchs to Noah and his sons, and to their descendants who were dispersed at Babel, and finally to the founders of various Mesopotamian city-states. Since Abraham came from Mesopotamia, it is likely that he observed a seven-day week, and thus his descendants, the nation of Israel, did too.

Apart from the biblical reason, there is no explanation of why people almost universally observe the seven-day week. Some people have suggested that the week is nearly one quarter of one month, effectively producing an intermediate time unit between the month and day. However, that is at best approximate. Furthermore, do we really need an additional time unit between the month and day, especially when the week does not evenly divide into the year or month any better than the day does? Others have suggested that there are seven heavenly bodies that move with respect to the stars (the sun, moon, Mercury, Venus, Mars, Jupiter, and Saturn) and that this is the reason why we observe the week. As we shall see, our names for the days of the week derive from pagan names of those heavenly bodies, and some people have used this as evidence for this explanation for the week. However, there is no historical record that this is the case. Rather, this explanation, like the other one, merely is conjecture.

There was at least one attempt to scrap the seven-day week. After their revolution, the French famously introduced the metric system, a decimal system of measurements. The Napoleonic conquest helped introduce the metric system across the continent of Europe. Less well known was the French attempt at measuring time using base 10. Their Revolutionary calendar consisted of a 360 day year with five or six extra holidays thrown in to bring their calendar back into reality. Each of their months was 30 days, consisting of three 10-day weeks. The day was divided into 10 decimal hours, each with 100 decimal minutes containing 100 decimal seconds. The French followed this time convention for about a decade. While most of the rest of the metric system established then was maintained, the new time standards proved to be so unpopular that Napoleon abolished them in 1805. The French Revolution was based upon humanism, and as such it was a rejection of both the monarchy and God. This changing of the week clearly was an attempt to thwart God's ordained time measurement. However, God's ordained time standard is ingrained even in the minds of unregenerate men.

The International Organization for Standardization (ISO [sic]) has attempted to standardize all sorts of measurements, including those for time. For instance, for a very long time people wrote out dates, such as October 4, 2015. However, over the past half-century it has become common to write dates in terms of numbers rather than by writing out the names of months. This may have coincided with the widespread use of computers, which require digital input. In the United States, the practice has been to mirror the manner in which we had written dates. That is, the digital equivalent of the month comes first, followed by the day and year. For instance, October 4, 2015 would be 10/4/15. However, elsewhere the practice generally has been to go from the smallest time unit to the largest. Hence, they would write October 4, 2015 as 4/10/15. Obviously, 10/4/15 and 4/10/15 ought not to refer to the same date.

The ISO standard is to go from the largest unit of time to the smallest unit of time. That is, October 4, 2015 would be written as 15/10/4, scrapping the US standard and reversing the standard from much of the rest of the world. Furthermore, there are other differences between the ISO standard and the manner in which we normally write dates. Instead of using a slash to separate years, months, and days, the standard is to use a hyphen instead. In the wake of the concern over Y2K at the turn of the millennium, the ISO standard actually is to write the four-digit date. (Early in computer use, storage capacity was at a premium, so years were truncated to two digits to save space—this is not a concern anymore.) Therefore, October 4, 2015 written as mandated by the ISO is 2015-10-04.

Obviously, writing dates in widely different ways leads to confusion, so such standardization is desirable. However, the ISO standard also is to treat Monday as the first day of the week. This clearly opposes God's standard of Sunday being the first day of the week and the Sabbath (Saturday) being the last day of the week. The ISO adopted the convention of the French beginning the week with Monday. This practice probably goes back to the Napoleonic era following the French Revolution. The early church observed Sunday as a day of worship (the Apostle John mentioned the Lord's Day in Rev. 1:10), but non-Jewish Christians never observed Saturday as a day of rest. Gradually, Christians began to view Sunday as a day of rest, though this is not a biblical teaching. As the French threw off God with their revolution, they decided to treat Sunday as a day of rest from work, but as an entirely secular exercise arising from man's reason. The French humanists then decided that it was best to begin the week with the first day of work.

The Names of the Days

The only day named in the Bible is the Sabbath. The other days of the week are merely numbered and never named. The Hebrew noun *Sabbath* (שַׁבָּת) derives from the verbal root שׁבת, which means "to cease" or "to rest." Clearly, this emphasizes the purpose of the Sabbath in the Old Testament Law.

Where do we get the names of the week? The Romans named the seven days for their gods that corresponded to the seven moving heavenly bodies. One can see some of those names in the names of the days of the week in the Romance languages. For example, in French Monday is *lundi*, Tuesday is *mardi*, Wednesday is *mercredi*, and Thursday is *jeudi*. In order, these refer to the moon (Luna), Mars, Mercury, and Jupiter. In English, the names of the days come from the names of Anglo-Saxon and Norse gods. There is some correspondence between Germanic and Roman gods, and so our days of the week amount to Anglo-Saxon versions of the Latin names. The names Saturday, Sunday, and Monday come from the names Saturn, sun, and moon, but technically the name of Saturn is a calque from Latin. Tuesday derives from Týr's Day, Týr being the Norse god of war. Týr is equivalent to the Roman god of war, Mars. Wednesday comes from Wōden's Day, Wōden being a variation of Odin. Thor's Day gives rise to Thursday, because Thor generally is thought of as being the German equivalent to Jupiter. This match is not perfect, because in Roman mythology Jupiter was the king of the gods, but Thor was not the king of the gods in Norse mythology. This equivalence probably comes from Thor being the most prominent Norse deity. Friday comes from Frige's Day, Frige being the Germanic goddess roughly equivalent to the Roman goddess Venus.

Starting with Saturday, the names of the days in order and in terms of the Roman god names are Saturn, the sun, the moon, Mars, Mercury, Jupiter, and Venus. Why are they arranged in this order? The arrangement generally is attributed to the ancient Egyptians. However, the arrangement is based upon astrology, and the ancient Egyptians learned much of their astrology from the Babylonians, so it probably originated in Mesopotamia. Ancient cosmologies had the fastest moving objects closest to the earth. Therefore, the seven heavenly bodies were, in order of increasing distance from the earth, the moon, Mercury, Venus, the sun, Mars, Jupiter, and Saturn. With the exception that the orbit of Mercury is farther from us than Venus' orbit, this is the correct ordering of these bodies in terms of increasing distance from the earth.

It was believed that each hour of each day was ruled by one of these seven moving objects. The name of a day came from the object that ruled the first hour of that day. The ruling went in order (generally, with one exception) of *decreasing* distance from earth. Starting with Saturday, Saturn ruled the first hour, Jupiter the second hour, Mars the third hour, the sun the fourth hour, Venus the fifth hour, Mercury the sixth hour, and the moon the seventh hour. After this, the ruling objects repeated throughout the remainder of the day and continued uninterrupted through each succeeding day. Therefore, Saturn also ruled the eighth, 15th, and 22nd hours of Saturday. The 23rd and 24th hours of Saturday were ruled by Jupiter and Mars, respectively. The next hour was ruled by the sun, and since this was the first hour of the next day, the day following Saturday is named Sunday. This pattern continues throughout the rest of the week, giving the ordering of the days' names. Notice that at the end of Friday, the first hour of the next day is named in honor of Saturn, thus repeating the cycle.

The Hebrew Festivals

There were three major feasts and associated days that the ancient Hebrews were to observe. Leviticus 23 lists these festivals and related observances. The weekly observance of the Sabbath is mentioned first (Lev. 23:3). Mentioned second is Passover (פֶּסַח), which was to be at twilight on the 14th day of the first month (Lev. 23:5). This month is called *Aviv* (Exod. 13:3–4; 23:15; Deut. 16:10). *Aviv* is the first month of the ceremonial calendar instituted with the first Passover in Egypt (Exod. 12:2). The month of *Aviv*, and hence the beginning of the ceremonial year, is near the vernal equinox. This is in contrast to the civil year, which begins six months later, near the autumnal equinox. Apparently, the ancient Hebrews continued to observe New Year (*Rosh Hashanah*) on the civil calendar, and it still is observed as such today.

Since Passover begins at twilight on the 14th day, and days in Hebrew reckoning begin at sundown, Passover coincides with the beginning of the 15th day. Being a lunar calendar, there always is a full moon on the 15th day of the month. This was providential, because the Hebrews departed Egypt during the night (Exod. 12:11, 29–32, 42; 13:3). Nighttime travel is difficult without artificial light. However, in most cases the full moon provides an adequate amount of light to travel safely.

Passover kicks off the Feast of Unleavened Bread, which was observed for seven days between the 15th and 22nd days of the month of *Aviv* (Lev. 23:6; Exod. 12:15–20; 13:3–10; Num. 28:16–25; Deut. 16:3–8). During this time there was to be no leaven in households, so people were to eat unleavened bread during this week. There were prescribed sacrifices for both Passover and the Feast of Unleavened Bread.

The Bible mentions two distinct sacrifices (wave offerings) associated with First Fruits. The first of these took place during the Feast of Unleavened Bread, on which occasion a sheaf of immature barley was offered (Lev. 23:9–14; Deut. 26:1–11). Much of Israel has a Mediterranean climate, with warm, dry summers and cool, wet winters. In such a climate, grain crops, such as the important grain barley, generally grew throughout the winter and ripened with the coming of spring. In fact, the Hebrew word אָבִיב (*abib*) refers to the stage of growth of grain when the seeds have reached full size, but have not yet dried. The word *abib* is used in Exodus 9:31 to describe the condition of the barley crop during the seventh plague of hail in Egypt. Hence, the name of the month of *Aviv* almost certainly comes from the association with the ripening of barley at that time.

It is believed that before the adoption of the Metonic cycle the maturity of the barley crop dictated whether to insert an intercalary month prior to the month of *Aviv*. A sheaf of grain from the upcoming harvest was given to the Lord as a wave offering symbolically representing the entire grain crop, along with prescribed animal sacrifices, grain offering, and drink offering. This was done on the day following the Sabbath that occurred during the week of the Feast of Unleavened Bread (Lev. 23:11).

It is not coincidental that Jesus rose from the dead the on the very day that the First Fruits were sacrificed. In refuting those who deny the resurrection of the dead, the Apostle Paul likened Jesus to the first fruit of the resurrection (1 Cor. 15:20–23). Jesus had compared His own death to that of a kernel of grain that must fall to the ground to be planted to bring an abundant harvest (John 12:24). The Apostle Paul expounded upon that analogy (1 Cor. 15:35–38) by comparing our bodies to seeds that must be planted to bring new life. The sacrifice of the First Fruits was at the very beginning of the harvest, but the abundance of the harvest came weeks later. In similar manner, Jesus' Resurrection was the very beginning, but we anticipate the future harvest of the resurrection of the saints.

The harvest itself was commemorated by the Feast of Weeks, or Harvest, the second feast required by the Law (Lev. 23:15–21; Deut. 6:9–12). The Feast of Weeks was seven weeks and one day after the sacrifice of the First Fruits. Since this is 50 days, this feast often is called *Pentecost*, from the Greek word for 50. Because the sacrifice of First Fruits always was on a Sunday, Pentecost always was on a Monday. Pentecost was after the completion of the grain harvest, so it amounted to a thanksgiving for God's provision. As with the other feasts and observances, Pentecost required a prescribed offering. Pentecost is most familiar to Christians as the time when the Holy Spirit descended on the Apostles (Acts 2:1–4).

The third prescribed feast was in the seventh month, near the time of the autumnal equinox, six months after the month of *Aviv*. The first day of the seventh month was marked with the blast of a shofar (Lev. 23:23–24; Num. 29:1–6). There was to be no work, and there was a prescribed offering. This day coincides with *Rosh Hashanah* (the "head of the year"). The tenth day of the seventh month was the Day of Atonement (*Yom Kippur*; Lev. 23:26–32; Num. 29:7–11), which also had a prescribed sacrifice.

The actual feast was the Feast of Booths, or Feast of Tabernacles, a seven-day feast that commenced on the 15th day of the seventh month (Lev. 23:33–43; Num. 29:12–39). As with the Feast of Unleavened Bread, which also began on the 15th day of its month, the Feast of Booths began with a full moon. The Feast of Booths was a time of thanksgiving for the autumn harvest, primarily of olives and grapes. Combined with this was the commemoration of the Exodus by living in booths constructed for this purpose (Lev. 23:42–43). The full moon closest to the autumnal equinox is called the harvest moon. Normally, the moon's rising is about 50 minutes later each successive night, but the harvest moon's rise is about 20 minutes later each night. This provides extra light shortly after sunset, adequate light to assist in harvesting crops, hence the name. This also was very favorable for the ancient Hebrews as they socialized in their booths, as the moon's light provided light to see well enough so that lamps may have not been necessary.

The Names of the Months

As with the days of the week, in the Bible most of the months are not named, but instead are mentioned in terms of their number. We have already seen that the Passover was explicitly stated to be in the month of *Aviv* (אָבִיב). There are only three other Hebrew names of the months mentioned in the Old Testament: *Ziv* (זִו, the second month; 1 Kings 6:1, 37), *Ethanim* (אֵתָנִים, the seventh month; 1 Kings 8:2), and *Bul* (בּוּל the eighth month; 1 Kings 6:38). The Hebrews eventually abandoned their Hebrew names of the months for Babylonian names, probably during the Babylonian captivity. These are the names used today in the Hebrew calendar, but the only scriptural references to any of those names are Nehemiah 2:1 and Esther 3:7, where the Babylonian name *Nisan* (נִיסָן) is used instead of *Aviv*.

Excursus: How Did Ussher Arrive at a Date of 4004 BC?

Archbishop James Ussher very famously fixed the date of creation in the year 4004 BC. More specifically, according to Ussher, the creation began on October 23 that year. Most people assume that Ussher reached his conclusions solely by adding together Old Testament genealogies. But the process was far more complicated than that. When Ussher first published his findings in 1650, there were already many such published chronologies. Even the famous astronomer Johannes Kepler had published his chronology several decades before Ussher. In short, assembling biblical chronologies was a very popular pursuit in Europe at the time, and this work attracted the attention of the most gifted minds. The fact that these various chronologies reached slightly different dates for the creation event indicates that there was some level of uncertainty involved. On the other hand, the fact that these many attempts approximately agree that the creation was about 4000 BC is a good indication that the Bible strongly indicates that the creation was about 6,000 years ago.

Constructing a chronology was not just a matter of adding together the biblical genealogies. Other passages not necessarily devoted to chronology contain subtle historical clues as well, and these had to be incorporated. Also, one must make many assumptions with regard to how the genealogies and other time clues are assembled. For instance, when the duration of time, in years, is stated for the lifetime of a patriarch or for a king's rule, there obviously is some rounding off of time. One must make assumptions about how to evaluate this. This alone easily could introduce 25 (or more) years of uncertainty as to the exact date of creation.

Furthermore, contrary to common misconception, Ussher and others did not rely solely upon the Bible's chronologies.[2] The Bible's internal chronology represents a good *relative* chronology, but the timeline is broken shortly after the return to the land after the Babylonian captivity, for there is a four-century gap between the Old and New Testaments. To establish an absolute chronology, one must fix the biblical chronology to a known date. This is done by correlating late Old Testament events with dates known from secular chronologies. Ussher amassed an impressive library of documents for this. However, much more information is available today that Ussher did not have access to. An important date in fixing the Bible's timeline is the fall of Jerusalem and the destruction of the first Temple. Ussher fixed the date of the first Temple's destruction in 588 BC, but today that event is dated in 586 BC. Obviously, a two-year difference 25 centuries ago is rather trivial.

Here we must note that astronomical considerations play a role in determining conventional dates in Middle Eastern history. Of particular importance is the total solar eclipse of June 15, 763 BC. For 150 years it has been widely accepted that this eclipse is the one mentioned in the Assyrian eponym lists as occurring in the year of *Bur-Sagale*. The phrase "solar eclipse" does not appear in the text. Indeed, there may not have been an appropriate term for a solar eclipse as we have today. The phrase appearing in the Assyrian record is *shamash* ("sun") *akallu* ("bent," "twisted," "crooked," "distorted," or "obscured"). This has been interpreted as referring to a total solar eclipse. The eclipse was short of being total at both Assur and Nineveh, the major Assyrian cities. However, the eclipse reference, if that is what it is, does not specify where the eclipse was observed. The eclipse was total in the northern part of Assyria, so certainly Assyrian officials would have known of the eclipse. This eclipse has been used to date the ninth year of the reign of Ashur-dan III. By correlating this occurrence with other events noted in the Assyrian eponym lists, we are able to determine the date of events that intersect with the Bible's history (see footnote 9 in Chapter 1 of this book for more detail). This has allowed us to fix the dates for most of the significant events in the Old Testament, including the fall of Jerusalem in 586 BC.

---

[2] For that matter, biblical events represent only a small part of Ussher's Annals. The vast majority of the events that he dated were non-biblical.

Notwithstanding this minor variance, using the date of 588 BC for the fall of Jerusalem, Ussher found that the creation was approximately 4000 BC. How did Ussher establish such a precise date of October 23, 4004 BC for creation? According to Jewish tradition, the Creation Week began with *Rosh Hashanah*. If one accepts that the ancient Hebrews were the recipients of ancient accurate record keeping, then this is a possibility. We know from the creation account that the creation began on a Sunday, the first day of the week. Therefore, only years in which *Rosh Hashanah* was on a Sunday would suffice. Furthermore, *Rosh Hashanah* must be the first day of the first month on the Hebrew civil calendar. Ussher reasoned that the only date that met these stringent criteria within the range of possible years of creation according to his calculation was October 23, 4004 BC.

CHAPTER 5

Biblical Teachings about and Warnings against Astrology

Astrology and astronomy are similar sounding words, but they have very different meanings. Both words derive from the Greek word *aster*, meaning star. The word *astronomy* comes from the union of this Greek word to the Greek word *nomos*, meaning law. Hence, astronomy literally means "star law." Thus, the word *astronomy* is not a bad choice for the name of the science, or systematized study, of the stars. On the other hand, the word *astrology* combines *aster* with another Greek word, *logos*, meaning word. But we have broadened the meaning of *logos* to be systematized knowledge. Some sciences, such as biology and geology, use the same root, so it is easy to think that astrology is a science too. However, it is not. Astrology is the belief that the positions of the sun, the moon, and the planets affect our lives and our destinies. Since astrology is more of a belief system than a science, it qualifies as a religion or a cult.

What the Bible Says about Astrology

More specifically, astrology is an ancient pagan religion. As such, it is not compatible with Christianity. The Bible contains only a few mentions of astrology, but where it does address astrology, it could not be clearer. The Old Testament Law probably focused on other pagan practices more than it did on astrology, because the ancient Israelites were far more affected by Canaanite culture than by other, more distant cultures. The ancient Canaanites do not appear to have been involved in astrology. The source of astrology as we know it is from the Babylonians, and their influence on ancient Israel was late, probably after the recovery of Hezekiah from his illness and the destruction of Sennacherib's army, the king of Assyria, about 700 BC. That left a little over a century of Babylonian influence before the captivity of the Jewish people in

Babylon. After the Babylonian captivity, the Jews largely were cured of their problem of embracing paganism, so astrology was not a major issue. However, though the Bible does not directly address astrology, as a pagan religion, involvement with astrology would at the very least violate the "great and first commandment" (Matt. 22:37). This commandment is the core of the Christian's expected complete devotion of heart, soul, and mind to God, so astrology is incompatible with Christianity.

One passage that might refer to astrology is Deuteronomy 4:19, which reads,

> And beware lest you raise your eyes to heaven, and when you see the sun and the moon and stars, all the host of heaven, you be drawn away and bow down to them and serve them, things that the LORD your God has allotted to all the peoples under the whole heaven.

However, this verse appears to refer explicitly to the worship of heavenly bodies rather than to the practice of consulting them for guidance, as is the custom of astrology. The book of Deuteronomy is the second giving of the Law, upon the imminent entering of the Promised Land after 40 years of wilderness wandering. The Law previously had been given at Mt. Sinai, but the generation that heard and experienced that had passed on. The new generation that was to inherit the Promised Land needed to receive the Law anew. In Deuteronomy, Moses reiterated the Ten Commandments (Deut. 4:13), relaying God's commission to Moses to teach His commands (Deut. 4:14). Deuteronomy 4:15–19 concentrates on the first two commandments. It explicitly commands the Israelites not to carve and worship images (Deut. 4:16–18). Many of the cultures that surrounded the Promised Land worshiped idols—as well as the sun, moon, and stars—as gods. Within this cultural context and the context of the biblical passage, Deuteronomy 4:19 quite literally is warning against bowing down to and worshipping heavenly bodies as the ancient Israelites' pagan neighbors were doing. This probably is the reason why the Genesis 1 creation account does not mention the sun and moon by name, referring to them instead as the greater and lesser lights. Besides being an historical narrative of the creation, the creation account also served as a polemic against the deities of the pagans that surrounded ancient Israel (see Chapter 1). In any case, it does not appear that Deuteronomy 4:19 is a specific warning against astrology, except in an oblique way.

Deuteronomy 17:2–7 details the procedure of judging and punishing a person who engaged in pagan idolatry. As verse 3 states, this includes one who "has gone and served other gods and worshiped them, or the sun or the moon or any of the host of heaven, which I have forbidden."

As with Deuteronomy 4:19, this verse condemns the worshipping of heavenly bodies along with worshipping any other false gods. Pagan cultures that surrounded the ancient Hebrews worshiped various deities, including the heavenly bodies. However, this is a bit different from astrology, which does not necessarily treat the heavenly bodies as deities. It is likely that some forms of ancient astrology viewed the various heavenly bodies as gods, but astrology today is not practiced that way. At the very best, this would be an indirect reference to astrology, at least as we know it today.

Much clearer is Isaiah 47:13–14:

> You are wearied with your many counsels: let them stand forth and save you, those who divide the heavens, who gaze at the stars, who at the new moons make known what shall come upon you. Behold, they are like stubble; the fire consumes them; they cannot deliver themselves from the power of the flame. No coal for warming oneself is this, no fire to sit before!

The entire chapter of Isaiah 47 is a pronouncement against Babylon. As previously mentioned, Babylon was the place of the origin of astrology, so this pronouncement against those "who gaze at the stars, who at the new moons make known what shall come upon you" clearly is a reference to the astrological practices of ancient Babylon. Equally clear is God's displeasure with their astrology. Notably, Isaiah 47:13 reads a bit differently in the King James Version. Rather than "those who divide the heavens," the King James reads, "the astrologers." Instead of "who gaze at the stars," the King James uses "stargazers." For "who at the new moons," the King James has "the monthly prognosticators."

Besides the use of the word *astrologers* in Isaiah 47:13, *astrologers* or *astrologer* is used eight other times in the King James Version, all in the book of Daniel. However, the word appearing in the original language is different in Daniel than the single occurrence in Isaiah. Notably in Isaiah, the phrase הַחֹזִים בַּכּוֹכָבִים literally means "star seers," which loosely translates into English as "astrologers." However, in Daniel the word אַשָּׁף (ʾašāp) is used. This word is difficult to translate, which is why it varies from one English translation to another. In all eight instances, it is used in conjunction with one to three other terms, thus lumping them all into a single category. For instance, Daniel 1:20 describes the progress of Daniel, Hananiah, Mishael, and Azariah in their education in Babylon. The King James Version says that the king "found them ten times better than all the magicians and astrologers that were in all his realm." But the English Standard Version renders this as the king "found them ten times better than all the magicians and enchanters that were in all his kingdom." The word translated "astrologers" or "enchanters" is the same Hebrew word noted above, 'ašāp. The uncertainty in the meaning of this word is the reason for this variation. Given that this was Babylon and that Babylon was the source of astrology, it is probable that translating this word as "astrologers" is warranted.

The word appears again in Daniel 2:2, where Nebuchadnezzar demands an interpretation of his first dream. The King James Version says that the king assembled "the magicians, and the astrologers, and the sorcerers, and the Chaldeans." In the English Standard Version, the assembled were "magicians, the enchanters, the sorcerers, and Chaldeans." The two English versions agree on three of the four groups mentioned, but 'ašāp is translated differently as in Daniel 1:20. Not much is known about the Chaldeans referred to here. One possibility is that this is not a reference to the Chaldean people in general, but rather to those schooled in the more ancient Chaldean learning. These people were highly regarded as very wise advisors. In subsequent verses (verses 4, 5, and 10) the Chaldeans acted as the spokesmen for all the wise men of Babylon, probably because of their highly exalted position.

In Daniel 2:10, the Chaldeans protested that the king's demand both to reveal the dream and its meaning was unreasonable, as no ruler had ever asked such a thing. Here the list of advisors being asked is pared down from the original four groups to three, with the sorcerers being omitted. The translations of the remaining three terms in the King James Version and the English Standard Version are identical to those appearing in Daniel 2:2. Because of the structure of the sentence, this is the one time that the words used for the groups of advisors are in the singular.

In Daniel 2:27, Daniel himself addressed the king. He noted that it was impossible for any man to achieve the feat that Nebuchadnezzar had proposed. More specifically, in the King James Version, Daniel noted that it was not possible for "the wise men, the astrologers, the magicians, [or] the soothsayers" to do this. In the English Standard Version, the list is "wise men, enchanters, magicians, or astrologers." The enumeration of four groups of people of Daniel 2:10 omits the sorcerers and Chaldeans of Daniel 2:2, but they are replaced by "wise men" and "soothsayers" (King James Version) or "astrologers" (English Standard Version). This portion of Daniel is in Aramaic, but the word for "astrologer" here is the same as the Hebrew word above.

Daniel 4 is the account of Nebuchadnezzar's second dream. In Daniel 4:6, Nebuchadnezzar summoned all the wise men of Babylon to interpret his dream, and Daniel 4:7 lists the wise men as they are listed in Daniel 2:10, except that "soothsayers" are the fourth group named in the King James Version, with the English Standard Version listing "astrologers" rather than "soothsayers," as it does in Daniel 2:27.

Moving forward in time, Daniel 5 details the incident of the handwriting on the wall during the reign of Belshazzar. In the King James Version, the word *astrologers* appears three times, in verses 7, 11, and 15. In Daniel 5:11, astrologers are listed with three other groups identically to how they are listed in Daniel 4:7. In Daniel 5:7, the list is the same, except that magicians are omitted. Finally, in Daniel 5:15, astrologers are mentioned only in conjunction with wise men.

Verse	Title (King James Version)			
Daniel 1:20	Magicians	Astrologers		
Daniel 2:2	Magicians	Astrologers	Sorcerers	Chaldeans
Daniel 2:10	Magician	Astrologer	Chaldean	
Daniel 2:27	Wise men	Astrologers	Magicians	Soothsayers
Daniel 4:7	Magicians	Astrologers	Chaldeans	Soothsayers
Daniel 5:7	Astrologers	Chaldeans	Soothsayers	
Daniel 5:11	Magicians	Astrologers	Chaldeans	Soothsayers
Daniel 5;15	Wise men	Astrologers		

The table summarizes the way in which the words *astrologer* or *astrologers* are listed in the book of Daniel. We may conclude several things about these mentions of astrologers. First, note that astrologers are listed with at least one other group of people in each instance. In the eight instances, the list is duplicated exactly only once, in Daniel 4:7 and 5:11. One ought not to infer too much from this. That is, one ought not to conclude that each instance is an exhaustive list. Rather, the lists are representative of various groups of men collectively considered sages. The differences probably amount to nothing more than stylistic variations, and the same large group of advisors probably is intended in each case.

Second, in none of these instances are astrologers singled out for condemnation. Rather, each appears to be an account of events that happened. In each of chapters 2, 4, and 5, the sages were unable to interpret the events, two dreams of Nebuchadnezzar and the handwriting on the wall. The inability of the sages to interpret these things is not condemned. Instead, God was honored by His ability to make these matters known (Dan. 2:28; 4:18; 5:17–28).

Third, we cannot say for certain that astrologers, as we know them today, are intended here, but they probably were included. The men referred to here were highly educated and trusted advisors. We must not forget that Daniel, Hananiah, Mishael, and Azariah were picked to be educated to serve in this corps (Dan. 1:3–7). These counselors advised the Babylonian kings in all manner of governmental matters, both domestic and foreign. The advisors probably spoke several languages and could hence serve as interpreters. They were educated in all that was known of the time. Besides subjects that we would approve of today, their education included instruction in the pagan religions indigenous to the Babylonian Empire, as well as occult practices—including astrology—associated with ancient Babylon. Hence, the corps of advisors were a mix of highly trained professional diplomats and pagan priests.

At the conclusion of their education, Daniel, Hananiah, Mishael, and Azariah had far more wisdom and understanding than their peers (Dan. 1:20). Being fully educated in pagan Babylonian knowledge undoubtedly was a challenge for devout followers of the true God, but Daniel, Hananiah, Mishael, and Azariah never wavered in their commitment to God (cf. Dan. 1:8–19; 3:1–29; 6:1–28). While many of the wise men of Babylon used occult methods in their advice, it is clear that Daniel and his associates did not.

Fourth, while the occult practices of the government advisors are not condemned in the book of Daniel, it does not follow that their occult practices, such as astrology, were alright. To the contrary, these occult practices, including astrology, were false religions that directly contradict the first commandment.

As previously mentioned, astrology, or even worshipping heavenly bodies, is not mentioned much in the Old Testament. This likely is due to the lack of local influence of cultures that did. However, worshipping heavenly bodies is mentioned late in the pre-exilic period. According to 2 Kings 21:1–17, Manasseh renewed pagan worship in Judah after his father largely had eradicated it. Verses 3–5 specifically mention him worshipping the host of heaven (2 Chron. 33:3–5). However, 2 Kings 23:1–20 later details how Manasseh's grandson Josiah once again removed pagan worship. Included in this removal was the worship of the sun, moon, the constellations, and all the host of heaven (2 Kings 23:5).

This all occurred in the 7th century BC, after the Northern Kingdom of Israel went into Assyrian captivity, but before the fall of Judah to the Babylonians. During this time, Babylon was in ascendency, and so Babylonian influence doubtless was increasing in Judah. The worship of heavenly bodies apparently was something new in Judah, so this practice may have been inspired by the

Babylonians. Still, astrology practiced as today does not involve worship of heavenly bodies as much as it is a belief that heavenly bodies influence our lives. In this sense, it may not be proper to equate the pagan practice of worshiping heavenly bodies with astrology. However, astrology undoubtedly has changed over the years, so this may have been the form of astrology that existed more than 25 centuries ago.

The Way Astrology Is Supposed to Work

As the earth orbits the sun each year, the sun appears to move through the stars once. The path through the stars that the sun appears to follow is called the *ecliptic*. The ecliptic is the plane of the earth's orbit. A 12° wide band centered on the ecliptic is called the *zodiac*. There are 12 constellations along the zodiac, so the sun is in each of these constellations approximately for one-twelfth of the year, or about one month. The constellation that the sun is in when one is born is that person's sun sign, or simply sign. For instance, when astrologers say that a particular person is a Libra, that means that the sun was in the constellation Libra when they were born. Due to influences that are never explained, all people of a particular sign are supposed to share common characteristics.

A problem arises right away. Astrology as it exists today was codified during the time of the Roman Empire, about 2,000 years ago. Due to precession of the equinoxes, the sun signs have shifted by about one position since then, but astrologers have failed to update their tabulation of sun signs to keep them synchronized with the real world. For instance, astrologers say that people born between September 23 and October 22 are Libras, but the sun actually is in Virgo during most of those dates. When confronted with this, most astrologers simply assert that while the stars may have shifted, the underlying signs that influence the affairs of people have not. That is, they separate the signs from the constellations. That probably is why astrologers seem never to use the word *constellation*, but rather use the word *sign* instead.

Astrologers use horoscopes to practice their craft. A horoscope is a diagram representing the positions of the sun, moon, planets, and astrological aspects at a particular time. For instance, a natal horoscope represents the situation when a person is born. To construct a natal horoscope, an astrologer must know the location, date, and time of a person's birth. Astrologers divide the zodiac into 12 houses. The first house is just below the eastern horizon and so is in the process of rising. The remaining 11 houses are sequentially numbered going eastward. Since the zodiac also is divided into 12 signs, each house is identified with a particular zodiacal sign. Suppose that at a particular location Aries is in the first house. Two hours later, Taurus will be in the first house, and so forth throughout the day.

Astrologers assign four primary angles. The most important angle is the first house. This is called the ascendant, because it is in the process of rising. The first house is called the east angle. The second most important angle is the tenth house, or north angle. This also is known as the midheaven, because it is located highest in the sky. Third most important is the seventh house, or west angle. It also is known as the descendent, because it is in the process of setting. The fourth most important angle is the fourth house, *imum coeli*, or south angle. This is the house that is lowest below the horizon. The ascendant, along with a person's sun sign, is supposed to be significant in a person's life. The other three primary angles play lesser roles.

The locations of the sun, moon, and planets within zodiacal signs and houses supposedly play their roles. However, more significant are the aspects of these bodies. An aspect is the relative angle between two objects. Conjunction occurs when the angle between two objects is 0°. Opposition is when the angle is 180°. Other aspects are square (90°), trine (120°), sextile (60°), sesquisquare (135°), and quincunx (150°). Some astrologers consider more minor aspects. Planets are at these relative angles only infrequently, but being close is considered good enough, though the power of an aspect wanes with departure from its exact angle. Astrologers express the amount of discrepancy from the exact angle by an orb, expressed in degrees from the exact angle. For example, if two planets, say Mars and Venus, are in conjunction, that will have some significance for a particular person, but the significance will be different if Mars and Venus are in opposition. Furthermore, influence of either will be lessened when the orb is, say, 4°.

The exact procedures and interpretations vary among astrologers. In essence, the interpretation of a horoscope amounts to a Rorschach test. While two astrologers may produce very similar horoscopes, their readings may be markedly different. That is why some celebrities employ several astrologers— because the astrologers generally will not give the same advice.

This raises a very good point. The sort of horoscopes that one sees in newspapers and magazines are not genuine horoscopes. These popular-level horoscopes lump all people with birthdays within a given month into one class. By lumping people together without regard to where and when (year, date, and time) they were born makes it impossible to compute a natal horoscope.

Without a natal horoscope, serious astrologers would consider it hopeless to produce a meaningful horoscope. Constructing and interpreting a natal and daily horoscope is time consuming. Consequently, the services of a professional astrologer are out of the price range of most people. However, affluent people sometimes waste their money on astrologers. One such astrologer was Carroll Righter, who was known as the "astrologer to the stars," because many of his clients were Hollywood actresses. Also influential in Hollywood was Joyce Jillson, and better known to the general public was Jeane Dixon.

While these astrologers offered their services to those who could afford them, the three also were responsible for daily horoscopes published in newspapers. These astrologers certainly knew that their daily columns were not serious horoscopes, so why did they produce them or at least give their approval for others to write them? There are several possible answers. One obvious answer is that newspapers paid for the horoscopes that they published, so this offered a steady income. Another possibility is that the astrologers thought that their daily horoscopes offered at least a little value to their readers. The rationale may have been that the value was commensurate with the cost, when one considers the relatively low cost of a newspaper and the fact that very few people purchase a newspaper solely for the horoscope. Still another possibility is that while the daily horoscopes did not offer what the astrologers considered real astrological advice, they did keep the astrologers' names in circulation and hence in peoples' minds. If people later decided that they wanted to consult an astrologer, this name recognition would have led potential customers to the astrologers. Therefore, while daily newspaper horoscopes may not be serious, they amount to a gateway to the occult.

The Status of Astrology Today

Despite the clear conflict between Christianity and astrology, astrology remained popular in the West throughout the modern Christian era. This probably came about because of the syncretism of the early church, as epitomized by the Roman Catholic Church. Even Johannes Kepler, who was devout and may truly have been a Christian, cast horoscopes for a living, at least part-time. Shortly after Kepler's lifetime, science as we know it developed and astrology was relegated to the cult that it is. However, during this time William Lilly published his oxymoronic book *Christian Astrology* in 1647.

Nevertheless, astrology continued to be popular in other forms. For instance, the practice of bloodletting was abandoned just two centuries ago

(George Washington probably died from the practice in 1799). The principle behind the practice of bloodletting was the doctrine of humors. According to this doctrine, the body contains four fluids, or humors: black bile, yellow bile, phlegm, and blood. An imbalance due to an excess or deficit in one of the humors was thought to be the cause of illness. Therefore, the remedy was to rid the body of one of the excess fluids by strategically letting fluid (blood) out of the body by way of an incision. How did one determine what the excess fluid was and hence precisely where on the body to let blood? This was determined by consulting a horoscope.

While bloodletting disappeared two centuries ago, "planting by the signs" still is practiced by millions of people today. *The Old Farmer's Almanac* was founded in 1792 and is the oldest continually published periodical in the United States. There are similar competing publications, such as the *Farmer's Almanac*. All of these almanacs contain many interesting tidbits, such as anecdotes, recipes, and gardening tips. Almanacs offer weather forecasts each year. Supposedly, these weather forecasts are made from secret formulae that rely upon planetary positions, qualifying them as astrological in nature.

Furthermore, such almanacs offer advice on when it is the best time to do certain things. For instance, the almanacs recommend cutting timber, cutting wooden shingles and shakes, and setting fence posts in the dark of the moon (from full moon to third quarter). The light of the moon (from new to first quarter) is recommended for planting aboveground crops, grafting trees, and for transplanting. You should dehorn and castrate animals during the moon's waning phases, but weaning should not be done then. According to this lore, certain crops should be planted during certain lunar phases, while planting of particular crops should be avoided during other phases. In addition, when the moon is in certain zodiacal constellations, conditions are favorable for planting certain crops, but planting should be avoided when the moon is in other constellations. Most almanacs will not specifically state which constellation is meant. Rather, each of the 12 zodiacal constellations is assigned a part of the human body, and recommendations are stated in terms of what part of the body "the signs" are in. For instance, Virgo is assigned to the bowels, so the two days or so each month when the moon is in Virgo is referred to as "the signs are in the bowels."

Obviously, this has an astrological basis, but many people today who use such instructions as their guide to gardening fail to recognize this. Instead, they reason that such lore is wisdom learned over a long time and passed down through the ages. More likely, someone promulgated a theory of when best to do some activity based upon what seemed astrologically sound to them, tried it, achieved success, and assumed that the astrological timing was the key to their success. The truth is, any seeds that are properly planted and cared for usually will grow successfully. The only way properly to test a hypothesis such as this would be to rigorously and repeatedly test the predictions accompanied by quantitative measurement of success. It is doubtful that this has ever been done.

Conclusion

For Christians who think that there may be something to astrology, it certainly is a bad idea to read horoscopes. Is it appropriate for Christians who do not believe in the power of astrology to read them? Since astrology is occultism, any dabbling in astrology amounts to an invitation to the occult. Aside from the potential danger of opening the door, however small, to the occult, why would a Christian want to read horoscopes? For the Christian, the clearest refutation of astrology is that it is a false religion. As such, it distracts from the proper worship of the true and living God. It also clearly violates the great and first commandment. That ought to be reason enough to avoid astrology.

PART 2

Astronomical Anomalies in the Bible: What Scripture Says about Unusual Astronomical Events

As Chapter 4 described, the astronomical bodies provide the basis for time measurement (Gen. 1:14). What we see in the sky changes throughout the night and throughout the year. This is due to the regular motion that astronomical bodies follow, which in turn is the result of God's current sustaining of this world (Col. 1:17; Heb. 1:3). If the motions of this world were not regular, then it would be chaotic, and astronomical bodies would be useless for the purpose of measuring time.

This regularity also makes science possible. If the world were in a constant state of flux, then one would have no confidence that an experiment repeated in more than one location and/or more than one time would have the same result. This is a deep philosophical issue. Most people assume that the regularity of the world is a given, that the world just naturally behaves this way. But why should the world be thus? This question is summed up nicely in a quote attributed to Albert Einstein: "The most incomprehensible thing about the universe is that it is comprehensible." Indeed, it is astonishing that the world is knowable. Unless, of course, there is a Creator. It is no accident that science as we know it developed in Protestant Europe. The Reformation brought about a renewed interest in the study of the world, such as astronomy, because it brings glory and honor to the Creator. If God, who is a God of order, created the world, then one might expect His creation to reflect God's attribute of being knowable. That was the working hypothesis that made science possible.

On the other hand, there are some extraordinary astronomical events recorded in Scripture, things that appear to fall outside the bounds of how things normally operate. Many people approach these accounts with the attitude that these events must be interpreted in terms of how the world generally works.

But if God created the world and ordained the order that now exists in the world today, is it not a small matter for God to intervene in His creation to alter, at least for a brief time, the manner in which the world works? To insist otherwise is to limit God's omnipotence and sovereignty. Thus, Christians need not be embarrassed or afraid to invoke miracles when miracles are clearly indicated by the biblical text.

In Part 2, we will explore the remarkable account of Joshua's long day and the unusual phenomenon of Hezekiah's sundial. Various authors have attempted naturalistic explanations for these events, but they seem to fall short of what the passages require. When we have eliminated the physically possible, it would seem to leave only the seemingly impossible, but God is pretty good at doing the impossible. We also shall briefly investigate the text of Matthew 2 to see what possible explanations could account for the star that led the magi to worship Jesus, the recently born king of the Jews. We shall see that the naturalistic explanations offered for the star of Bethlehem fall short too. That leads once again to the consideration of a less conventional answer to the question of the Christmas star's identity.

Also in Part 2, we will look at some clues from Scripture that allow us possibly to fix the date of the Crucifixion and Resurrection of Jesus Christ astronomically. In conjunction with this, we can investigate the likelihood of some claims commonly made today about astronomical events allegedly concurrent with the Crucifixion.

Finally, Part 2 concludes with a discussion of the astronomical events that are found in prophetic passages in the Bible. This inevitably will take us into the potentially murky waters of eschatology. There are many different opinions about end-time events. We can be certain of only one thing—they all can't be right. Hopefully, no one will be upset by what I have written on the subject, but instead all will come away with renewed interest in the very interesting, but perplexing, astronomical aspects of prophetic passages.

CHAPTER 6

The Unusual Days of Joshua and Hezekiah

Joshua's Long Day

Joshua 10:12–14 is the remarkable account of what most people refer to as "Joshua's long day." These three verses have spawned all sorts of explanations and interpretations. A straightforward reading of the text indicates that in order for the Israelites to conclude a large and very successful battle against their adversaries, the Lord stopped the sun in the sky, thus considerably extending the length of the day. Therefore, the orthodox explanation is that the earth stopped rotating for about a day, whereupon the earth began rotating once again.

Many critics seize upon Joshua 10:12–14 as proof that the Bible is geocentric, teaching that the earth does not orbit the sun. This charge is discussed more fully in Chapter 13 of this book. Suffice it to say here that there is nothing inconsistent in believing that the sun is the center of the solar system while speaking of the sun moving across the sky. Astronomers today speak of the *apparent* motion of the sun and other astronomical bodies across the sky as the earth rotates on its axis each day, but this does not indicate that they are geocentrists. When astronomers mention sunrise and sunset, that does not indicate that their cosmology is geocentric, but rather they are using the language of how this motion appears. The language of appearance equally applies to the Bible as well. Joshua 10:12–14 is not describing cosmology, but rather it describes what people at the battle experienced.

For those who believe that the Bible is inspired and that the God of the Bible is omnipotent, the account of Joshua 10:12–14 clearly describes a miracle. However, we ought not to gloss over the physical problems with this miracle, which is the reason why so many people—including, unfortunately, some truly born-again people—seek other explanations. The earth is a spinning body. Spinning bodies possess kinetic energy, and the earth's kinetic energy is considerable. To bring the earth's rotation to a stop requires removing that kinetic energy. This could be done by applying a torque to the spinning earth, acting as a brake. Application of a braking torque takes time. Furthermore, a brake must convert the kinetic energy to some other form of energy, usually heat. What was the nature of this braking torque? What happened to the kinetic energy removed? Furthermore, after the battle was completed, the earth began spinning again, which would have required time and an accelerating torque to reinvigorate the earth with kinetic energy. What produced this accelerating torque, and what was the source of the required kinetic energy?

But another physical principle is involved. Moving objects possess a quantity called momentum. Dissipating the earth's rotation required loss of angular momentum that would have transferred to the object providing the braking torque. In a similar manner, returning the earth's spin afterward would have required the addition of angular momentum, presumably at the expense of the object responsible for applying the accelerating torque to the earth. Furthermore, objects on the surface of the earth are spinning around the earth's axis of rotation. People on the battlefield that day were moving nearly 900 miles per hour toward the east. When the earth suddenly stopped rotating, those people and everything else on the surface of the earth would have continued moving eastward, so they would have appeared to suddenly begin moving nearly 900 miles per hour across the earth's surface. Even a gradual slowing in the earth's rotation would have produced a noticeable "push" to the east, similar to the outward push that we feel as a car quickly rounds a curve. In a comparable manner, when the earth's spin was returned, people and objects on the earth's surface would experience a "push" with respect to the earth's surface, this time to the west. To avoid this, God would have had to have applied the same change to people and objects on the earth's surface that He applied to the earth itself.

These are the sort of physical reasons that have led some people who profess belief in Scripture to argue for some explanation for Joshua 10:12–14 *other* than what the plain reading of the text suggests. These same people appear to have no problem with the miracles of the Virgin Birth or the Resurrection of Jesus Christ. But just as our understanding of physics alone would seem to eliminate the miracle of Joshua 10:12–14, our understanding of how conception occurs would seem to eliminate the possibility that a man could be born without the participation of human sperm. Nor would our understanding of biology today allow for the body of a man who had been dead for three days to come back to life. Yet we understand by the miraculous intervention of our Creator God that Jesus was virgin born and that He rose from the dead. In a similar manner, we ought not to doubt that God in His infinite power and wisdom could accomplish the miracle described in Joshua 10:12–14. Yet many professing Christians do reject this miracle. Before discussing different attempts to explain away the miracle, we ought to examine the lead-up to the miracle.

The Background

The book of Joshua records the beginning of the conquest of Canaan. Chapter 6 tells of the fall of the powerful city of Jericho. Chapter 7 describes the initial defeat at Ai, followed by the ultimate victory over Ai in chapter 8. After hearing of the conquest of those two cities, in chapter 9 the leaders of Gibeon, the next logical stronghold to attack, sent emissaries to Israel to forge an alliance via deception. Chapter 10 opens with Adoni-zedek, king of Jerusalem, establishing an alliance against Gibeon with four other Amorite kings of cities south and west of Gibeon. These kings realized that they likely were the next targets of Israel, so by striking against Israel's new ally, they thought that they could deliver a defeat to Israel. The leaders of Gibeon sent word to Joshua while he and Israel were camped at Gilgal near Jericho, requesting Israel's assistance against the Amorite alliance (Josh. 10:6). Joshua led his men into battle (verse 7) after the Lord assured Joshua of victory (verse 8). Joshua's attack commenced after an overnight march from Gilgal (verse 9). This was a distance of nearly 20 miles, which would have required marching all night. Joshua 10:10 states that the Lord threw the Amorites into a panic, struck them, and chased the Amorites. The path of their retreat was down the road by Beth-horon, west-northwest from Gibeon, though this initial retreat does not lead directly back to their cities. From there, the Amorites turned southwestward all the way to Azekah (verse 11). This turn would have taken most of the men back toward their home cities. According to verse 11, the Lord threw down large hailstones from heaven, killing more Amorites with these hailstones than the Israelites killed with the sword. It is important to the discussion to note that this retreat path was more than 25 miles.

It was at this point that the writer of Joshua interjected the account of Joshua's long day (verses 12–14), and concluded with the statement that Joshua and Israel returned to Gilgal (verse 15). It is easy to read the events of Joshua 10 as sequential, with the events of the rest of the chapter (Josh. 10:16–43) following the return to Gilgal noted in Joshua 10:15. However, there are reasons to believe that the description of the events found in the verses immediately following Joshua 10:15 is a recapitulation of events that actually happened before the return to Gilgal mentioned in Joshua 10:15. Many more recent translations place the verb in verse 16 in the pluperfect tense, suggesting this understanding (ancient Hebrew verbs do not have tense, so translating the verb as a pluperfect reflects the understanding of the translator, who draws his translational conclusions from the literary context). Joshua 10:43, the final verse of the chapter, records that Joshua and Israel returned to Gilgal, using language identical to Joshua 10:15. While it is possible that Joshua and Israel could have returned to Gilgal twice, it makes more sense that the return happened once, with two independent mentions of that return, with either recounting of the return to Gilgal following the two descriptions of events that preceded the return.

There are contextual reasons for believing that the events of Joshua 10:16 and following were concurrent with the events of verses 1–14. Verse 16 states that the five kings of the alliance hid in the cave of Makkedah. When Joshua was informed of where the five kings were hidden in verse 17, Joshua responded in verse 18 by giving the order to place large stones in front of the entrance of the cave and to leave a few guards. Joshua admonished his men in verse 19 not to be concerned with the five kings right then, but to continue to press the battle in a mopping up operation to prevent the rear guard of the Amorite forces from returning to their cities. The location of Makkedah is uncertain, but it probably was south of Azekah. Thus, the likely location of Makkedah is at least 50 miles from Gilgal by the route that the Israelite forces had taken. The implication of returning to the camp at Gilgal was for rest and recuperation. It would have taken at least a couple of days to have returned to Gilgal. After a few days of rest, it would have taken two days to return to Makkedah. Thus, at least a week would have transpired between the first and second attack, if there were two returns to Gilgal. It is unlikely that the five kings would have gone into hiding between the two battles; it would have made more sense for them to seek the relative safety of their walled cities. The rear guard would have had more than ample time to have retreated back to the safety of their cities as well. Therefore, it makes more sense that the return to Gilgal happened once—after the other events recorded in Joshua 10—not twice.

Once the remnant of the Amorite forces managed to make it back to their cities (verse 20), the Israelite forces camped at Makkedah. The purpose of this camp probably was to recuperate from the very exhausting battle that they had just completed. After marching all night, and fighting and chasing the enemy throughout an incredibly long day, the troops would have required extensive rest. It would have made no sense for them to have traveled all the way back to Gilgal for this. Thus, the rest at Makkedah likely was the evening after the very long day. In Joshua 10:22–27, Joshua gave the command to remove the five kings from the cave, whereupon Joshua killed the kings. This event likely did not happen on the same day as the battle, because with the five kings imprisoned in the cave, there was no need for haste. Furthermore, the kings were executed and their bodies hung on trees for the remainder of the day, whereupon their bodies were removed and interred in the cave at sunset. It is unlikely that this was the same as the long day, suggesting that this day at the very least was the day following. The remainder of the chapter recounts the defeat and destruction of several cities in the region, including three of the five cities of the alliance. The implication is that each city fell rapidly after a very brief siege, but still it would have required, at minimum, several days to accomplish the entire conquest of this region. It is likely that the entire campaign took up to a month or two. The later conquest of the north recorded in Joshua 11 would have taken considerable time. This suggests that the return to Gilgal at the end of Joshua 10 was for an extended time to rest and prepare for the next campaign.

Various Explanations

What are some of the interpretations of Joshua's long day? One suggestion is that a total solar eclipse cooled and refreshed the troops during a tough battle. The major problem with this explanation is that a total solar eclipse lasts at most a few minutes, so the relief would have been momentary, and one hardly could have expanded this to refer to nearly a day. Furthermore, a solar eclipse happens only when the moon and sun are in the same part of the sky, but the account of the long day explicitly describes the sun and moon being in two different locations. Furthermore, the moon is not even visible for a few days around a solar eclipse.

Related to this explanation, some have suggested that great cloud cover arose or some other darkening came up to give relief from the sun's heat. Substantiation of this supposedly comes from the language that Joshua used in verse 12: he literally commanded that the sun be silenced. Since the sun does not emit an audible sound, some reason that this must refer to the sun's heat. But Joshua's wording "Sun, stand still at Gibeon" (שֶׁמֶשׁ בְּגִבְעוֹן דּוֹם) is an idiom, which means "to bring to rest." Furthermore, verse 13 uses a non-idiomatic expression to indicate that the sun stopped. Hence these sorts of attempts seriously oppose the language used. Another problem with both the solar eclipse and cloud cover theories is that these mechanisms would have benefited both sides of the battle, but the text clearly indicates that only Israel had the advantage.

Some have suggested a sort of strange refraction effect occurred so that the sun continued to illuminate the battlefield, even though the sun had set. This in itself would have required some sort of miracle in the sense that it would have been a radical departure from how the world normally operates, so it is not clear how this is physically preferable to a stop in the earth's rotation.

A more liberal approach is to hypothesize that with God's help the troops performed so well that it just *seemed* to them like the day's light was extended. That is, *there was no miracle*, but it seemed as if one occurred. But this would have required that the Israelite forces chased the Amorites at least 25 miles, assuming that the pursuit ended at Azekah. If the battle continued at least as far as Makkedah and beyond, as appears likely from the text, then the attack would have been carried out over nearly 50 miles. This would have to have been accomplished in just a normal day, and that after a 20-mile overnight march! No matter how well the battle progressed, this is not possible during a single, normal day. One could respond that God endowed the Israelites with extreme endurance and speed to accomplish this. However, this would have extended to the Amorites too, for without this miraculous help as well, the much faster Israelites would have rapidly overtaken the Amorites, and so the chase would not have been nearly as long as recorded. It is not clear why this miracle is preferable to the miracle clearly indicated by a straightforward reading of the text.

An even more liberal explanation is that there was no miracle at all, nor was there the implication of anything unusual about the battle. Rather, the description of Joshua 10:12–14 merely is poetic hyperbole. A comparison is drawn between Joshua 10:12–14 and the description of the stars fighting against Sisera in the song of Deborah and Barack found in Judges 5:20. Of course, the song of Deborah and Barack clearly is labeled as such (Judges 5:1). Songs, being poetry, often contain hyperbole. Is there a poetic element to this account in Joshua? Yes, for the latter part of verse 12 and the first part of verse 13 are written in a form of poetry.

"Sun, stand still at Gibeon,
  and moon, in the Valley of Aijalon."
And the sun stood still, and the moon stopped,
  until the nation took vengeance on their enemies.

Notice the parallel between lines one and two. Also notice the parallelism contained in line three. Line three also repeats what lines one and two said. Hence, there are poetic elements here. However, those who take these words merely as poetic hyperbole willingly ignore the rest of verses 13 and 14. The second half of verse 13 rhetorically asks if this extraordinary day is not written in the Book of Jashar.[1] Furthermore, the second part of verse 13 goes on to use very literal, non-metaphorical language to state that the sun stopped in the middle of heaven and delayed setting for about a whole day. Appeals to poetic form cannot trump what appears to be a straightforward statement. Additionally, the statement of verse 14 that there has been no day before or since like that one makes sense only if there was something truly remarkable about that day. The biblical author taking poetic license hardly qualifies for that description.

What Actually Happened

Many people picture this miracle happening late in the day, and Joshua expressing his desire that the light of day be extended shortly before when the day would have otherwise ended. Joshua commanded the sun to stand still at Gibeon and the moon to stand still in the Valley of Aijalon. The Valley of Aijalon is west of Gibeon, down the valley in the direction that the Amorites initially retreated. Since the Amorites were making war against Gibeon, it makes sense that the battle began at Gibeon. If the battle were late in the day, then the sun would have appeared in the west. However, it was the moon that was described as being in the westward direction, not the sun. One could surmise that when Joshua gave the command, he and his troops were between Gibeon and the Valley of Aijalon. That would place the sun over Gibeon in the east, which would require the time to be early morning. Why would Joshua have been concerned about having time to complete the battle before nightfall, if it were still early in the day?

---

[1] The Book of Jashar is mentioned one other time in the Bible, 2 Samuel 1:18. Nothing of the Book of Jashar survives, so we do not know its content. From the two mentions in the Old Testament, we may infer that it was a historical book that existed at that time but now is lost. This appeal to the extrabiblical literature suggests that this account is an historical narrative.

A more plausible explanation is that Joshua gave the command at midday while he was at Gibeon. This would have placed the sun high in the sky, essentially over his location at Gibeon. In fact, this is the sense of verse 13, for it states that the sun stopped in the "midst of heaven" and did not hurry to set for about a whole day. The "midst of heaven" (בַּחֲצִי הַשָּׁמַיִם) would be the middle of heaven, suggesting a time near midday, or noon. Perhaps the rout of the Amorites had commenced, and Joshua already realized by noon that extra time was required to complete the defeat of the Amorites. What of the mention of the moon? If the moon were at or just past third-quarter phase, then at noon it would have appeared low in the western sky over the Valley of Aijalon. In that manner, the sun would have been stayed over Gibeon while the moon remained over the Valley of Aijalon.

From this it is clear to anyone who takes the text seriously that the earth's rotation stopped for the better part of a day so that the battle could be completed. It is not clear from the description of Joshua 10:13 that the sun did not hurry to go down for about a whole day if this refers to most of the daylight portion of day (12 hours) or to an entire day (24 hours). If the former, the sun shone for at least 18 hours, but it could have shone for as long as 30 hours. At the conclusion of the battle, the earth began to rotate again. There is no avoiding the conclusion that this was a miracle. Some have argued that the language of Joshua 10:12–14 required that the moon also stopped in its orbit around the earth. However, that is not necessary. In the course of 24 hours, the moon moves 13 to 14° westward. If the earth's rotation stopped, the moon would have crept slightly toward the sun's position, its motion solely being that of its revolution around the earth. That change hardly would have been noticeable to people who were continually on the move and engaged in serious combat—for all practical purposes, the moon would have remained motionless in the sky.

Does this miracle have an astronomical explanation? That is, could some astronomical event have taken place to cause this miracle? Some people have suggested that a large body passed near the earth at the time of Joshua's battle at Gibeon. In this manner, these people hope that some sort of tidal interaction could have de-spun the earth and then re-spun it. However, many of the physical problems already mentioned would apply anyway. For instance, even if such an interaction were to have changed the earth's spin, it would have done little to keep people from flying across the earth's surface, as previously noted. Furthermore, such a close pass of a massive body would have raised unbelievably high tides on the earth. There is no record, either written or geological, of such an event. There is no known astronomical phenomenon that could have produced this event in a natural way.

Hezekiah's Ten Steps

The account of Hezekiah's sundial is found in three passages in the Bible. Most people consider 2 Kings 20 to be the most complete description. In verse 1 of 2 Kings 20, Hezekiah became very sick, and the prophet Isaiah came and told him to get his affairs in order, for he was going to die. Verses 2–3 record that Hezekiah turned his face to the wall, that he wept, and that he prayed earnestly for healing. Isaiah had not traveled far when God called him back (verse 4). God instructed Isaiah to tell Hezekiah that He had heard his prayer and seen his tears and that Hezekiah would be healed in three days (verse 5). Furthermore Hezekiah would live another 15 years (verse 6). Isaiah prescribed a poultice to place on Hezekiah's boil (verse 7). Hezekiah then asked for a sign that his healing indeed would happen (verse 8). God asked whether Hezekiah would prefer that the sun's shadow moved forwards or backwards ten steps (verse 9). Hezekiah reasoned that going forwards would be relatively easy, because that is the direction that the sun would travel anyway (verse 10). Therefore, Isaiah prayed to God, and the sun moved back ten steps (verse 11). The steps here are described as "the steps of Ahaz." This is probably because Ahaz previously had built the structure in question. The remainder of 2 Kings 20 records the visitation of envoys from Babylon and events from the rest of Hezekiah's reign. There it states that the envoys were sent bearing gifts from the king of Babylon, because the king had heard that Hezekiah had been sick.

A parallel passage is 2 Chronicles 32:24. However, this verse records only bare details—that Hezekiah became sick to the point of death, that he prayed to the Lord, that God answered Hezekiah's prayer, and that God gave him a sign. The nature of the sign is not described, but this certainly must have been the sign of 2 Kings 20. After detailing other actions of Hezekiah, 2 Chronicles 32:31 briefly mentions the visit of the Babylonian envoys, offering one more detail. The princes of Babylon had sent the envoys to enquire about the sign. Of course, there is no reason why the envoys were not sent to do both—enquire about the sign *and* Hezekiah's health. Note that 2 Chronicles 32 states that the princes (plural) of Babylon had sent the envoys, while 2 Kings 20 says that the king of Babylon (singular) sent them. Probably, there was bureaucracy involved, so that one could describe the mission as ordered by either a single official or several officials.

The third passage, Isaiah 38, offers almost the exact same account of 2 Kings 20. The largest difference is that Isaiah 38:9–20 includes a song of Hezekiah not recorded in 2 Kings 20. Furthermore, Isaiah 38:7–8 has Isaiah telling of the sign without a mention of Hezekiah's request for a sign. However, after Hezekiah's song, two more details are listed. Isaiah 38:21 contains Isaiah's prescribed treatment, a detail left out of the earlier part of the chapter. Furthermore, Isaiah 38:22 states that Hezekiah asked for a sign. In some more modern English translations, the verbs in these verses are in the pluperfect tense, indicating a past action. We have already seen that the narrative of Joshua 10 is not strictly in chronological order, and this is another example of this practice. However, it is easy to place the events in chronological order because of the required progression of events. Furthermore, 2 Kings 20 includes both of these details in chronological order. At any rate, all three treatments present unity in their narratives.

What were the steps of Ahaz? We are not sure. They likely were literal steps on a stairway in Hezekiah's palace. The language certainly seems to indicate that. However, some have suggested that these steps were markings on a sundial, for the word used here (מַעֲלָה) could mean that. It is not clear that sundials were in use in that part of the world at that time. But if it is so, this is the earliest mention in the Bible of the measurement of time other than in days, months, or years. Certainly the New Testament has many references to the Roman time convention of 12 hours during the day and 12 hours during the night. It is most likely that the standard of 12 hours developed in Babylon. Since there is explicit mention of Babylon by this point in the Bible, it is possible that this form of time reckoning was at the very least familiar to the Hebrews in Hezekiah's lifetime. However, even if that were the case, this does not mean that this structure alluded to here was an actual sundial as we now know it. Rather, this could have been a stairway that happened to make a useful reference for approximately measuring time during the day when the sun's shadow was visible. If a wall is properly oriented, the shadow of another object, say another tall wall, may move down and up the wall, as well as across the wall, throughout the day. A stairway along the wall can provide useful reference points during the shadow's journey. This often occurs by happenstance, but a good architect easily can design such a thing within a building's plan.

Some who argue that this was an actual sundial believe that the ten steps were 10° markings on a sundial. As with the division of day and night into 12 hours each, our current practice of subdividing a circle into 360° originated in Babylon. Ten degrees of the sun's motion corresponds to 40 minutes of time. If this is correct, then the miracle was the reversal of the sun's shadow by 40 minutes. On the other hand, if the ten steps were some other division of time, or if the steps were actual steps on a stairway, then the amount would have been far more, probably several hours. If the sun reversed only 10°, most people probably would not have noticed this, unless they were directly observing the sun's shadow at the time that it happened. Certainly, Hezekiah and his court would have been watching in anticipation, because Isaiah had made them aware of the impending miracle. However, recall that the Babylonians sent envoys to enquire about this sign. If they had heard by word of mouth of some sign in a distant, rather obscure kingdom, it is unlikely to have warranted much attention. But if they had witnessed this event themselves and then learned of why it had happened, that would have been important enough to have sent representatives. The Babylonians might not have directly observed the reversal of the sun's shadow had it been a modest 40 minutes, but if the reversal were a few hours, it would have been obvious to nearly everyone outdoors at the time that the sign happened. Therefore, it seems unlikely that the steps were single degree marks on a conventional sundial.

However great the reversal in the sun's shadow was, how did God accomplish this? This sign would have required not only that the earth's rotation stop, as with Joshua's long day, but the earth's rotation must have reversed before resuming its normal motion again. The narrative suggests that the changes in the earth's rotation were very fast, requiring perhaps only a few seconds. The physical difficulties that applied to Joshua's long day also apply here. As with Joshua's long day, no astronomical event can explain this. Therefore, this must have been a miracle as well.

Has Science Confirmed These Miracles?

There is a marvelous little story about Joshua's long day that began circulating in the late 1960s and early 1970s, during the heyday of the Apollo program. According to the story, in preparation for the Apollo moon landings, a computer at NASA calculated the positions of the earth, moon, and other solar system bodies with great precision far into the past and future. This computer program produced a glitch in the 15th century BC, a glitch caused by solar system bodies not being in their correct positions, indicating that nearly a day was missing from time. An additional 40 minutes also was missing several centuries later, so that the total missing time was one full day. Supposedly, NASA scientists and engineers puzzled over this problem until one of them opened the Bible to Joshua 10:12–14 and 2 Kings 20:8–11. The NASA personnel supposedly came to realize that their missing day could be explained by addition of nearly a day at the time of Joshua and an additional 40 minutes at the time of Hezekiah, thus proving that these biblical events actually occurred.

This story was carried in a few newspapers at the time, but it enjoyed widespread circulation among Christian audiences, and it occasionally is heard even today. At the time, computers were a bit mysterious, and many people thought that computers could do almost anything. People generally don't think that now, but the story continues to circulate, primarily because it has been around for so long and has appeared in so many sources that it is easy to find references to the story. The person who appears to be responsible for this story is Harold Hill, president of the Curtis Engine Company in the late 1960s. Hill told the story a number of times before finally committing it to print in a 1974 book, *How to Live Like a King's Kid*. Hill's company had done contract work for NASA during the Apollo program, and Hill claimed that during this time he became aware of the computer confirmation of Joshua's long day through his work at NASA. In some versions of the story, Hill worked for NASA, but Hill never did. Furthermore, the sort of work that Hill's company did would not have placed him in a position to be privy to this sort of information. Hill claimed to have seen documentation of this supposed event as a NASA memo, but Hill could not produce the memo. Some people have speculated that if Hill actually did see such a memo, the memo may have been an April Fool's Day prank.

The problem with this story is that a computer could not find such a discrepancy in the positions of solar system bodies. In order for a discrepancy to be found, we must know exactly where things were at some time in the past prior to Joshua's long day, so that the calculated positions can be compared. A computer can calculate only where things ought to have been in the past; it cannot compute where things actually were in order to make a comparison. Computers can be used to determine how much the earth's rotation has slowed over the centuries due to the tidal interaction between the earth and moon. We can calculate into the past when and where total solar eclipses should have occurred on the earth absent the earth's slowing rotation. From historical records we know when and where certain eclipses were actually observed. From this discrepancy, we can measure how much the earth's rotation has slowed. However, in the case of Hill's computer story, such a comparison is not possible, because we do not have information as to the location of various solar system bodies at an earlier time.

This story is not new, but rather it is a modern retelling of an even older story. In the 1930s, Harry Rimmer made reference to how science had proved the missing day of Joshua, and this story continued to circulate within Christian circles for decades. Rimmer's mention of this may have been the origin of Hill's story. Rimmer based his statement upon an 1890 book by C. A. L. Totten, *Joshua's Long Day and the Dial of Ahaz, a Scientific Vindication and "A Midnight Cry."* Totten did a very elaborate computation of the date of the battle of Gibeon with respect to the creation of the world. He reasoned that the battle was on the 24th day of the fourth month of the Hebrew civil calendar in the 2,555th year after the Creation. This was the 933,285th day since creation. From this, Totten determined that this day was a Tuesday. Next, Totten calculated backwards in time from June 17, 1890 to the battle of Gibeon. He concluded that the battle was 1,217,530 days previously, which was a Wednesday. Hence, there was a day missing. Of course, Totten's computation required very precise dates, something that most people today would find ludicrous. However, Totten managed to obtain some audience in the late 19th century. While most people today are not impressed with such an approach, apparently invoking a computer, as in the Hill story, is sufficient to convince some people. This story has been debunked many times, so it is a shame that it keeps being repeated.

CHAPTER 7

The "Christmas Star" of Matthew 2

It seems that everyone is familiar with the "Christmas star," because it is such a memorable part of the Christmas story. Most people recall that three kings, or three wise men, traveled hundreds of miles by camel to visit and pay homage to a newborn baby Jesus. The most important portion of the story, at least as far as this book is concerned, is that a star led them the entire way. Unfortunately, the details of this story, like so much of the Christmas story, come more from Christmas songs, Christmas cards, and those countless Christmas pageants put on by children in bathrobes, than from the Bible. Consequently, many of the details of this particular part of the story are not actually found in Scripture but instead have crept into our common understanding. Let us sort through this tangle to see exactly what we know from the biblical account.

The Biblical Account of the Christmas Star

The wise men are discussed in only 12 verses, in Matthew 2:1–12. Who were these men? Matthew 2:1 calls them wise men, as found in most translations. However, some translations call them "magi," a word taken from the Greek word μάγοι (*magoi*). The ancient Greeks got this term from a Persian word referring to a class of highly educated advisors. The Persians considered them to be very wise, so "wise men" is a good translation. Much of their education would have been in Chaldean knowledge, and at least some of this knowledge likely would have included what we would call occult practices today. For instance, magic is considered part of the occult. We get the word *magic* from the same Greek root as that from which we get the word "magi." The Greek noun *magoi* in Matthew 2:1 is plural, but the singular form of this noun is used in Acts 8:9 and Acts 13:6 to refer to two different men who were magicians or sorcerers. The two magicians in Acts were depicted as evil, but the magi were not. Why the difference? The two magicians in the book of Acts were not Persian, nor had they gone through the extensive education of the Persian magi, and hence they were not wise. Rather, they had dedicated themselves entirely to the dark arts. Today we frequently call people who perform tricks "magicians," even though their performances do not rely upon spirits. Rather, they perform illusions, and many modern magicians prefer to be called "illusionists." So the word *magoi* can have either a good or bad connotation depending upon the context.

How many magi were there? The Bible does not say. Since the word is in the plural (in English the singular form of magi is magus), we know that there were more than one. The tradition that there were three of them probably comes from the three gifts—gold, frankincense, and myrrh—listed in Matthew 2:11. However, there is no reason why only two magi or more than three could have offered these gifts. Some people have attempted to spiritualize these gifts. But there probably was not more to them than the fact that they were compact, yet expensive, so they were obvious gifts to take on a long trip. The magi are never identified as royalty, so it is not proper to refer to them as kings, though such is how they are often depicted. This misconception may stem from their identification as trusted advisors among the Persians. Highly trusted advisors often were given much authority in government,[1] and this may have morphed into the legend that they were kings.

Where did the magi come from? One Christmas song suggests a land called Orientar (Orient are), but such a place is unknown to historians.[2] *Magi*, the word used to describe them, is of Persian origin, which suggests that they came from the region of Persia, far to the east of Israel. At the time of Christ's birth, Persia was part of the Parthian Empire. The western end of the Parthian Empire was the eastern border of the Roman Empire. The major cities of the Parthian empire were in modern day Iran and Iraq, so it is likely that they traveled at least from Mesopotamia along the Fertile Crescent. Matthew's Gospel does not reveal what animals, if any, they rode. If the magi rode rather than walked, they could have used camels, but they more likely would have employed horses, mules, or donkeys.

---

[1] For instance, Daniel and his three friends were educated in Chaldean knowledge (Dan. 1:3–4, 17–20), and they were given much responsibility in government (Dan. 2:46–49). Though this predates the Persian Empire and hence the use of the word "magi," Daniel was from the group of people that the word magi described.

[2] One cannot help but think of how the words of this song are commonly slurred together—especially by young children singing it for the first time!

Matthew 2:1 records that the magi arrived in Jerusalem sometime after Jesus was born. According to Matthew 2:2, the magi enquired in Jerusalem about the birth of the King of the Jews, for they had seen His star in the east, and had come to worship Him. The phrase, "star in the east" (cf. Matt. 2:9) as it quite literally is translated in the King James Version and the New American Standard Bible is problematic. This phrase can be understood three different ways. First, it could refer to where the magi were when they saw it, in the east, or east of Jerusalem, which would be where they lived. Second, it could mean that the magi saw the star in the eastern part of the sky. This raises the question in many peoples' minds why the magi proceeded westward when they saw the star in the east? Related to the second possibility is the third possibility, that the magi saw the star as it rose, which is how the English Standard Version and New International Version translate this phrase both times that it appears in Matthew 2. Due to the earth's rotation, most celestial objects rise in the east and set in the west each day. Therefore, this possibility suggests that the star first became visible as it rose. This interpretation enjoys wide support today, as evidenced by its appearance in the New International Version and the English Standard Version.

The account of the rest of the magi's time is Jerusalem is found in Matthew 2:3–8. Apparently, the magi did not directly ask King Herod about this matter. When Herod heard about the magi and their question, he was troubled, probably because he viewed the birth of a new king as a threat to his rule. Herod asked the scribes and chief priests where the Messiah was to be born, and they replied that from Micah 5:2 it was Bethlehem. Herod secretly met with the magi to enquire about when they had first seen the star. Herod then sent the magi to Bethlehem with instructions to search for the child and when they had found the child to return and inform him, so that he, too, could go and worship. Matthew 2:9 states that the magi departed for Bethlehem and that as they did, the star "went before them until it came to rest over the place where the child was." Matthew 2:10 tells us that, "When they saw the star, they rejoiced exceedingly with great joy."

When the magi went into the house, they saw the child and his mother Mary, they fell down and worshiped him, and they opened their treasures and offered him gifts. Finally, in a dream the magi were warned not to return to Herod, so they traveled home by a different route that avoided Jerusalem. Subsequently, Joseph also was warned in a dream to flee to Egypt (Matt. 2:13–15).

This was to avoid the slaughter of all the male children who were two years old or younger in Bethlehem and the surrounding region once Herod realized that the magi would not return (Matt. 2:16–18).

There are several observations that we can make. One is that there is no indication that the star led the magi on their long journey from the east to Jerusalem, as in the common conception. The description of the star leading, or preceding, the magi comes from the account of their relatively short journey from Jerusalem to Bethlehem. Bethlehem is only about six miles south of Jerusalem, so even with a slow walking pace, one can traverse that distance in just a few hours. This is in contrast to the many weeks or even months required for the magi to reach Jerusalem from their homes. There is no mention of the star similarly leading the magi on the much longer journey to Jerusalem. The common misconception that the star led the magi the entire way probably stems from convolving the two mentions of the star.

Indeed, the mention of the star again after the magi had left Jerusalem, the more detailed description of the star's behavior then, and the delighted reaction of the magi to seeing the star, all suggest that the magi may have not even seen the star throughout their long journey to Jerusalem. The magi's reference to the star in Matthew 2:2 is in the past tense, also implying that the star was not visible in Jerusalem, even to the magi. At any rate, the fact that no one in Jerusalem seemed to know what the magi were talking about suggests that the people in Jerusalem had not seen the star, or at the very least had failed to appreciate its significance. If the star had been exceedingly bright (per the common conception) and visible throughout the world, then it is very unlikely that the people of Jerusalem would have been ignorant of the star. Therefore, it is possible that the star either was not that bright or it was not widely visible.

Before moving on to possible identifications of the Christmas star, we probably ought to address the issue of timing. Most depictions of the first Christmas show the magi visiting the night of Jesus' birth, along with the shepherds. Luke 2:11 makes it very clear that the shepherds arrived the same night that Jesus was born. However, there are several reasons to believe that the magi arrived much later, possibly months or even a year later. First, Herod had all male children the age of two and under in Bethlehem killed (Matt. 2:16). Part of the two years may have reflected the time that that magi reported that they had first seen the star, indicating the Messiah's birth. It would have taken weeks or even months for the magi to reach Jerusalem, but certainly not two years. There probably was some time after the magi met with Herod before he realized that they were not returning. Furthermore, Herod may have included a safety factor in his order to ensure that he was able to eliminate any possible usurper to his throne. At any rate, it appears from this piece of information that the magi did not visit when Jesus was a newborn. Second, the Greek word for "child" (παιδίον) used in Matthew 2:11 does not refer to a newborn but rather to a toddler or a young child. Third, the Old Testament Law required an offering of dedication for the firstborn son 40 days after birth. The appropriate offering was a lamb and a dove or pigeon (Lev. 12:6), but if they could not afford that, they were to offer two doves or two pigeons (Lev. 12:8). According to Luke 2:22–24, Mary and Joseph offered the latter sacrifice. This indicated that they were too poor to afford the lamb, but if the magi had already visited with their expensive gifts, that would not have been the case. Therefore, the magi must have visited at least 40 days after Jesus' birth. Furthermore, the expensive gifts that the magi brought likely financed the flight to Egypt. If so, then this was within God's providential timing.

Suggested Identifications of the Star

There has not been a shortage of suggestions as to what the Christmas star was. As we saw in Chapter 3, the modern concept of what stars are is a bit different from how people in the past viewed stars. In ancient languages, the words for stars referred to any point-like sources of light in the sky. This excluded the sun and moon, since they clearly are not point-like. Today we understand that the sun is a star, but that concept would have been foreign to most cosmologies of the past. To the naked eye, the planets appear like stars in the sky, so they were classified as (wandering) stars. Even today, when most people happen to notice a planet in the sky, they think that it is a star. In the past, meteors were thought of as being stars, but today we realize that they are bits of material that burn up as they enter earth's upper atmosphere. Comets appear as hairy stars. While airplanes and spacecraft did not exist until recently, they too appear as stars in the night sky, so to the ancient understanding, these would qualify as stars too. Most of these objects as well as actual stars have been suggested as having been the Christmas star.

For instance, 400 years ago the famous German astronomer Johannes Kepler (1571–1630) observed a rare triple conjunction of the planets Jupiter and Saturn. Due to their orbital motions around the sun, the planets generally appear to move eastward with respect to the stars. Jupiter takes 12 years to complete one circuit through the stars, while Saturn takes 30 years. At 18 to 20 year intervals, Jupiter and Saturn are in the same part of the sky, an event that we call a conjunction. Usually, this is a single conjunction, in that Jupiter slides past Saturn once on its somewhat faster motion through the stars. However, every 13 months, the earth passes Jupiter as the earth moves much quicker on its smaller orbit around the sun. This causes Jupiter to appear to move backward, or retrograde (westward with respect to the star rather than eastward) for about four months.[3] Similarly, Saturn experiences retrograde motion every 12½ months, its retrograde motion lasting about 4½ months. Moving more quickly in its orbit and its orbit being only half the size of Saturn's, the distance that Jupiter appears to travel in its retrograde motion is greater than the apparent distance that Saturn travels. On rare occasions, the retrograde motions of Jupiter and Saturn coincide with their conjunction so that Jupiter passes Saturn three times, once just prior to retrograde motion, once during retrograde motion, and once after retrograde motion. Kepler observed one of these rare triple conjunctions. It is often erroneously reported that these events are far rarer than they actually are, taking seven or eight centuries to repeat. They are more common than that—there were two triple conjunctions of Jupiter and Saturn in the 20th century (1940–1941 and 1980–1981), though there will be none in the 21st century.

Kepler computed when triple conjunctions of Jupiter and Saturn occurred in the past, and he found that one happened in 7 BC. He noted that this was close to the date of Jesus' birth, so he suggested that this triple conjunction of Jupiter and Saturn was related to the Christmas star. This triple conjunction happened in the constellation Pisces, which supposedly had some important significance to Israel. This suggests an astrological connection. The magi, trained in the Chaldean arts and science, probably were well-versed in astrological lore. For decades this has been the staple of planetarium shows during the Christmas season. However, nearly all of these shows omit one very important element of Kepler's theory. Kepler did not claim that the triple conjunction itself was the star. Instead, he thought that the triple conjunction caused a nova, and that it was the nova that was the Christmas star.

---

[3] See Chapter 13 for a greater discussion of retrograde motion.

A nova is a star that experiences a very rapid, extremely large increase in brightness, followed by a more gradual decline. The star that brightens, what astronomers call the progenitor star, usually is too faint to be seen with the naked eye, so a nova that is bright enough to be seen with the naked eye briefly appears as a new star. In fact, the word *nova* comes from the Latin word for "new." Nearly a century ago, astronomers realized that some novae (the plural form of nova) increase in brightness far more than others, so they call these extremely rare events supernovae. The theoretical understanding and hence the origin of novae and supernovae are very different, but in ancient times there was no distinction made between the two. Besides Kepler, others have suggested that the Christmas star was a nova or supernova. Apparently, some early church fathers thought that the Christmas star was a nova. There are Chinese records of a nova in 5 BC. Some have suggested that this was the Christmas star, or that it was some other nova for which no record survives.

Still others have suggested that the Christmas star was a comet. For instance, Halley's Comet was brightly visible in 12 BC, though that date is a bit early. There probably were other comets for which no records survive that could have been the Christmas star. The appearance and motion of comets are very different from the other candidates for the Christmas star so far discussed, which might conform better to the brief description of the star from Matthew's gospel. However, comets generally have been interpreted as bad omens. For instance, apparitions of Halley's Comet have coincided with disastrous events, such as the Battle of Hastings in 1066. It is unlikely that a comet would have been interpreted as a good sign by anyone. Interestingly, some theological liberals have suggested that the next apparition of Halley's Comet after 12 BC was the inspiration of Matthew's account of the star. They suppose that Matthew wrote his gospel late, about the time of the destruction of Jerusalem and Herod's Temple in AD 70. The destruction was precipitated by a rebellion that began around AD 66. Halley's Comet hung over the city at that time, and some people later interpreted the comet's apparition as an omen of the coming destruction. According to this theory, Matthew's account of the star was pious fiction that Matthew worked into the story, inspired by the appearance of Halley's Comet in AD 66.

Additionally, some people have suggested that the Christmas star could have been a reference to a series of lunar occultations of stars or planets. An occultation is when an object passes in front of a second object, occulting, or blocking out, the second object. The moon is the most common occulting object, and we call these events lunar occultations. The moon's orbit is inclined a little more than 10° to our orbit around the sun, so the only objects that the moon can occult are those that lie within a band around the sky that is a little more than 10° wide. A few bright stars and all five naked-eye planets are within this band. Due to the 18.6-year nodal precession of the moon's orbit, there are seasons approximately 18.6 years apart where occultations of a particular object can take place. This cycle is long enough to make occultations relatively uncommon, but not nearly long enough to make them that rare. For this reason, this theory largely relies upon the timing and order of occultations to serve as the Christmas star. However, this very fact makes it very subjective to interpret, which seriously weakens the argument for it.

In recent years, much attention has been given to another conjunction explanation. In 2 BC, there was a very close conjunction of the planets Venus and Jupiter. This conjunction was so close that the two briefly would have merged into one to the naked eye. This may have been the only time that such an event happened in recorded history, prompting some people to identify this event as the Christmas star. However, there are several problems with this interpretation. First, the conjunction happened only very briefly (less than an hour), low in the western sky shortly after sunset, and then only for a very narrow range of longitude. This longitude did coincide with the Middle East, but this also required that the sky was reasonably clear for this to be observable. Second, this was as these objects were setting, not rising, and many people think that the words describing the magi seeing the star in the east refers to its rising. Third, one must creatively interpret the text of Matthew 2:9–10 where the magi saw the star again, because the close conjunction was long past. Fourth, and perhaps most damaging, this event happened after 4 BC, the well-established date of Herod's death.

Supporters of this theory have responses to many of these objections. For instance, they strongly support the thesis that Herod died in 1 BC rather than 4 BC, as will briefly be discussed in Chapter 8. However, very few, if any, historians have endorsed the later date of Herod's death. They further argue that the close conjunction of Venus and Jupiter was just the most spectacular of several conjunctions that took place about that time. They point out that there was another conjunction of Venus and Jupiter nine months earlier in 3 BC. But such conjunctions are not that rare, for they must happen roughly on an annual basis, and conjunctions a few months apart are not uncommon. For instance, there were two conjunctions of Venus and Jupiter in 2015, one in late June and the other in late October. The next conjunction was in late August 2016. Incidentally, the two summer conjunctions were very close, though the two planets did not appear to merge to the naked eye. Supporters of this theory emphasize that the planetary conjunctions of 3–2 BC happened in the constellation Leo (or lion), which allegedly has significance for Israel, or at the very least Judah, based upon Genesis 49:9 (cf. Rev. 5:5). However, recall the triple conjunction of Jupiter and Saturn in 7 BC was in the constellation Pisces, which supposedly had an association with Israel. Finally, in the interval 3–2 BC there was a triple conjunction of Jupiter and the bright star Regulus. The name Regulus means "royal" (we get the word *regal* from the same root), because Regulus is one of four royal stars (according to astrology, the other three royal stars are Aldebaran, Antares, and Fomalhaut). Jupiter comes into conjunction with Regulus roughly every 12 years, and many of those are triple conjunctions as well. Hence, conjunctions of Jupiter and Regulus are not that remarkable.

Proponents of this theory claim that the remarkably close conjunction of Venus and Jupiter and the other related phenomena must be interpreted. They argue that since the magi were well educated in astrology and astronomy (the two were mingled at the time) that they would have understood these things. However, it is conjectured how the magi would have interpreted these things. If one can see significance in a conjunction in Pisces while another sees significance in a conjunction in Leo, then this amounts to a Rorschach test. Most people agree that the Christmas star conforms to one of the purposes of the stars given in Genesis 1:14, to be "for signs," but a sign is effective only if it is clearly understood (for instance, a road sign written in a language unknown to the reader is of no help). If a sign requires tremendous interpretation, then its effectiveness is seriously compromised.

Then there is the matter of the description of the star in Matthew 2:9. The theory that the star was a conjunction of Venus and Jupiter would require that these planets be visible in the sky once again. The star is described as going before the magi and then standing over the place where the child was. However, in the few hours that it may have taken to travel southward from Jerusalem to Bethlehem, the planets would have moved westward, or to the right as one faces south as the magi would have been. The same problem exists for other known astronomical objects that we have previously discussed. In no way could one say that the star went before them. Nor could any normal astronomical body have stood over the place where Jesus was. Supporters of the various natural astronomical objects claim that if the object were low in the sky, it would appear as if it were over the town of Bethlehem. However, this is insulting to the magi. As skilled observers of the heavens, as is often claimed, then the magi had to know that from a particular vantage point, almost any astronomical body could appear to be over a particular location. This amounts to a triviality, and so does not constitute any sort of sign, and thus is hardly worthy of mention.

As it turns out, this latter objection applies to all natural astronomical explanations for the Christmas star. Perhaps the Christmas star was not a natural phenomenon.

A Non-Natural Explanation

Recall that in the languages of the Bible, stars are any point-like objects in the sky. Even artificial lights, such as those on aircraft at night, if they had existed back then, would qualify by this definition. Hence, many people have suggested the star that the magi saw was no ordinary star. Rather, it could have been a special light source sent by God to be a sign only for the magi. This light would have appeared as a normal star, but it could have been at an elevation of only hundreds or thousands of feet. It would have been geographically limited in its appearance, and thus would have limited the number of people who could have seen it. Such a light source could have appeared and disappeared as necessary, and it also could have appeared to move or hover in one spot. Thus, it could have fit the description of the Christmas star's behavior, something that purely natural objects could not do.

What might this object have been? There are several possibilities. One possibility is that it was a specially created object to serve this purpose for a limited time. Another possibility is that the light was supernaturally generated with no physical source. Most people who favor this explanation fall into one of two camps. One suggestion is that the star was the manifestation of an angel or angels. In speaking of creation, Job 38:7 says it was a time "when the morning stars sang together and all the sons of God shouted for joy."

Most commentators think that this refers to angels (cf. Job 1:6; 2:1). If so, then stars and angels are equated here, though, being a poetic expression, we may not want to take this too literally. Elsewhere, we see that Revelation 12:3–4 describes a dragon in heaven and his tail sweeping down "a third of the stars" to earth. Revelation 12:7 states,

> Now war arose in heaven, Michael and his angels fighting against the dragon. And the dragon and his angels fought back.

Revelation 12:9 continues,

> And the great dragon was thrown down, that ancient serpent, who is called the devil and Satan, the deceiver of the whole world—he was thrown down to the earth, and this angels were thrown down with him.

This is the same dragon from verse 3, who is now identified as Satan. He is accompanied by "his angels" who oppose Michael and his angels (verse 7). As is evident in the text, Michael's angels serve God, so these other angels must be angels who fell along with Satan. The wording of verse 9 of Satan falling to the earth is reminiscent of Luke 10:18 where Jesus said, "I saw Satan fall like lightning from heaven." Some see here an allusion to Isaiah 14:12, which reads,

> How you are fallen from heaven, O Day Star, son of Dawn! How you are cut down to the ground, you who laid the nations low!

This is from the English Standard Version. The King James Version, by contrast, reads "Lucifer" (a transliteration of the Latin word for "light bearer") rather than "Day Star." However, most modern translations of the Bible do not translate this as "Lucifer," but instead use "Day Star" or some equivalent. This verse comes from a taunt against the king of Babylon (Isa. 14:3–21), but since some people think that Luke 10:18 is an allusion to Isaiah 14:12, they think that Isaiah 14:12 has a dual referent, and that it has in view the devil as well. However, keep in mind that this idea is controversial, because many Old Testament scholars do not agree with this.[4]

At any rate, understanding that angels and stars have some equivalence in the biblical text, we are justified in surmising that the third of the stars of heaven that were swept down to earth in Revelation 12:4 were the angels who joined in Satan's rebellion, thus becoming fallen angels, or demons. Hence demons represent one-third of the angels that God created. All of this is to say that, at least in some contexts, angels can be thought of in terms of stars, and presumably so can demons. In 2 Corinthians 11:14 we see that "Satan disguises himself as an angel of light." This statement primarily is intended as a figurative one, as darkness is used as a metaphor for evil (cf. Luke 22:53; Acts 26:18; Eph. 6:12; Col. 1:13). However, there is no reason why this could not have a literal component. At any rate, it is clear, if angels can have some equivalence with

---

[4] The same can be said about Ezekiel 28:11–19. This passage is a condemnation of the king of Tyre that some people see as having a dual meaning referring to Satan. However, many Old Testament scholars dispute this.

stars, then angels can appear as stars. Therefore, many people have concluded that the Christmas star was an angelic phenomenon.It is interesting that the Jehovah's Witnesses teach that it was a demonic star rather than angelic source from God.[5]

Still others hypothesize that the Christmas star was the Shekinah, or the presence of the glory of God. The Shekinah sometimes appeared as a cloud during the day, but as light at night (Exod. 14:19–20; 40:34–38). Revelation 21:23 tells us that the New Jerusalem will have "no need of sun or moon to shine on it, for the glory of God gives it light, and its lamp is the Lamb." Jesus said, "I am the light of the World. Whoever follows me will not walk in darkness, but will have the light of life" (John 8:12). In his first epistle, the Apostle John wrote, "God is light, and in him is no darkness at all" (1 John 1:5). James 1:17 refers to God as "the Father of lights." Clearly, some of these passages use light in the figurative sense, but Revelation 21:23 does not. Moreover, there is no reason why even the figurative use of light cannot have a literal component. These verses have led many to conclude that God was the source of the light that He called into existence on Day One and continued to be that source until Day Four when He made the sun. This conjecture is in response to the question about the source of light on the earth prior to the creation of the sun, but it is a good conjecture and constitutes what probably is the best answer to that question. Similarly, this is a good answer to the question of the identity of the Christmas star.

Conclusion

Ultimately, does it really matter what the Christmas star was? Not really, for God did not see it necessary to reveal that to us. However, God has made us curious creatures, and so it is natural to ask about such things. It is okay to seek such answers, but we must remember to seek answers that are consistent with the biblical text.

Many Christians have found encouragement in the natural explanations for the Christmas star, such as the explanation involving the conjunctions during 3–2 BC. This encouragement apparently stems from the confirmation of things in the Bible, thus amounting to evidence that the Bible is true.

---

[5] It is interesting that the Jehovah's Witnesses teach that the Christmas star was a demonic star rather than angelic source from God. We must note, however, that the view of the Jehovah's Witnesses is completely unconnected to the identification of the star as an angel by some conservative Evangelical scholars.

However, this can work in the opposite direction as well. Skeptics consider the Bible to be a collection of tales. Therefore, it is just as easy for a skeptic to embrace naturalistic explanations for things found in the Bible. If a series of conjunctions occurred near the time of Jesus' birth, then that could have been woven into Matthew's gospel, and in the process robbed it of any special significance. Thus a naturalistic explanation confirms the skeptic's worldview, and he can find encouragement in the natural explanation too.

As followers of Christ, should we be afraid to invoke the miraculous? After all, a Virgin Birth is something that modern science has not observed, and hence would appear to be impossible. The same is true of the Resurrection. We do not seek natural explanations for either the Virgin Birth or the Resurrection of the Lord Jesus Christ. Rather, we accept them in faith as being the miracles that they were. Similarly, creation was miraculous. Not only is it ill-advised to seek natural explanations for creation, but it is impossible as well. This does not mean that everything in the Bible is beyond the normal operation of the world and hence must be miraculous. The Christmas star could have had a natural explanation. However, it is difficult to square Matthew's account of the star with known astronomical objects. Given that, we ought not to be afraid to seek out a miraculous explanation for the Christmas star.

CHAPTER 8

Astronomical Aspects of Good Friday and Resurrection Sunday

Determining Date of Resurrection Sunday

As discussed in Chapter 4, the ancient Hebrews used a lunisolar calendar. The first day that the thin crescent moon was visible in the western evening sky was the first day of the month. The following days were numbered sequentially throughout the month until the next thin crescent was visible in the western evening sky, thus ushering in a new month. Because the synodic period of the moon is approximately 29½ days, the months normally alternated between 29 and 30 days. The beginning of each month originally may have been observationally determined. However, with record keeping, one can learn to anticipate when the thin crescent in the western evening sky would be visible, even when inclement weather made direct observation impossible. Therefore, an algorithm was developed to establish a calendar. This was very useful, because it permitted people to know in advance when the months would begin, and when the various Hebrew festivals would be. This allowed people to plan ahead for the various Hebrew festivals. The algorithm used likely changed with time. Today we can use computers to calculate accurately when the first thin crescent would have been visible from Jerusalem at any time, thus allowing us to determine with much confidence when each month began during biblical times. More specifically, this permits us to determine when the dates of the Hebrew festivals, such as Passover, occurred in antiquity.

At Mt. Sinai, God instituted the observance of Passover to commemorate the ancient Hebrews' liberation from bondage in Egypt (Exod. 23:15; cf. 12:1–28; Lev. 23:5–8). The first Passover was the 15th day of the first month of the ceremonial calendar (the ceremonial calendar also was established at

Mt. Sinai). The original Passover, the night that God killed the firstborn in each household of the Egyptians, was the night of the 15th day on what was to become the ceremonial calendar, so each Passover is on the exact date that the original Passover miracle happened. Of course, this is the exact date only on a lunisolar calendar. We use the Gregorian calendar, which is a solar calendar, so the date of Passover moves about on our calendar.

All four Gospels record that Jesus was crucified at the time of Passover. For instance, Matthew 26:1–2 states that Jesus reminded His disciples that the Passover was only two days away, and that the Son of Man would be delivered up to be crucified. This is followed by Matthew 26:3–5, which informs us that the chief priests and scribes conspired to kill Jesus. In a parallel passage, Mark 14:1–2 tells us that two days before the Passover the chief priests and scribes conspired to take Jesus' life. Another parallel passage, Luke 22:1–2, merely states that the Passover was drawing near when the chief priests and scribes sought to kill Jesus. John 13:1 states that before Passover, Jesus knew that His time had come to depart this world to go to the Father.

We know that Jesus rose from the dead on Sunday, the first day of the week (Matt. 28:1–7; Mark 16:1–11; Luke 24:1–9; John 20:1). Since Jesus rose from the dead on the first day of the week, Sunday has special significance for Christians. If each year we are to observe the anniversary of the Resurrection, it makes sense to do so on a Sunday. However, which Sunday should this observance be? As we shall see, Passover began a few hours after the Crucifixion, which means that Jesus rose from the dead shortly after Passover. Therefore, it makes sense to observe Resurrection Sunday immediately following Passover. Note that unlike Resurrection Sunday, Passover can fall on any day of the week.

Passover is the 15th day of the first month of the ceremonial calendar. Since the ceremonial calendar is lunisolar, the 15th day of the month always falls on a full moon. The first month of the ceremonial calendar normally is around the time of the vernal equinox. To replicate these conditions, the algorithm for determining the date of Resurrection Sunday is the first Sunday after the full moon following the vernal equinox. Protestants and Roman Catholics observe Resurrection Sunday on the same date. Resurrection Sunday for the Eastern Orthodox Church usually coincides with the Western observance. However, the definition of terms within the algorithm used by the East and West are slightly different, so occasionally Resurrection Sunday for the Eastern Orthodox Church and Resurrection Sunday for the Roman Catholic Church and Protestant Church differ by a month. Similarly, Passover usually occurs a few days prior to Resurrection Sunday. However, the two are separated by a month occasionally. Because of the way that it is defined, Resurrection Sunday, just like Passover, moves about the Gregorian calendar. In the West, the earliest possible date of Resurrection Sunday is March 22, and the latest possible date is April 25.

Did the Crucifixion Happen on a Friday?

While all four Gospels agree that the Crucifixion was near Passover, certain details appear to present contradictions. The manner in which we handle these issues determines the day on which we think that the Crucifixion occurred. We generally commemorate the Crucifixion on the Friday before Resurrection Sunday, because John 19:31 states that the day after the Crucifixion was a Sabbath. For this reason, the Jews asked the Romans to kill Jesus and those crucified with him. It was against the Old Testament law for the body of a person executed on a tree to remain on the tree after sundown (Deut. 21:22–23). However, this does not appear to have been the Jews' motivation, for John 19:31 specifically states that it was concern over the Sabbath. Apparently, the Jews thought that leaving the bodies up on the Sabbath, which commenced at sundown, would defile the Sabbath. The verse further mentions that this Sabbath was exceptional, for it was a high day. Therefore, it would appear that the day after the Crucifixion not only was a Sabbath, it was Passover as well. It may be that the primary concern was impropriety of crucifixions continuing into Passover.

What could be done to ensure the rapid death of those crucified? People who were crucified, being suspended with their arms outstretched, found it difficult to breathe. They managed to breathe by pushing up with their legs and holding that position. Soon their legs tired, so they had to let their legs relax. This relaxation made it difficult to breathe, so they would push up again. However, if their legs were broken, they no longer could push upward, and they soon suffocated. According to John 19:31–37, the Roman soldiers broke the legs of the two thieves crucified with Jesus to hasten their deaths. However, Jesus was already dead, so the soldiers did not break his legs. This was in accordance with Scripture that none of His bones would be broken (Ps. 34:20).

All four Gospels relate the burial of Jesus (Matt. 27:57–61; Mark 15:42–47; Luke 23:50–56; John 19:38–42). Luke 23:56 clearly implies that the burial was accomplished prior to sundown, the beginning of the Sabbath. John 19:42 concurs in noting that the tomb was close to Calvary, making it easier to complete the burial prior to sundown. Preparation of a body for burial constituted work, and work was forbidden on the Sabbath, as well as on the first day of Passover (Exod. 12:16). Furthermore, to work on the Sabbath or Passover would have been inconsistent with the character of Joseph of Arimathea, who in Luke 23:50–51 was described as a "good and righteous" man. Hence, Jesus was buried late Friday, just prior to sundown.

However, this presents a problem, because Jesus was to be buried for three days. As we have seen, the Resurrection occurred early Sunday morning. With Hebrew reckoning of time, the Resurrection could have been any time after sundown on what we would call Saturday evening, though it could have been as late as shortly before sunrise on Sunday morning. Supposing that Jesus was buried just before sundown on Friday, it would appear that the minimum that He was in the grave would have been slightly more than 24 hours, but the maximum could not be much more than 36 hours. This is far short of three days as we understand temporal reckoning today.

There are several ways to resolve this dilemma. One approach is to note that ancient Hebrews counted any portion of a day as a full day. Hence, the small portion of Friday that Jesus was in the grave would have constituted one day. Similarly, since Jesus did not rise before sundown on the Sabbath, which ushered in Sunday, He was in the tomb on Sunday as well. With this reckoning, being in the grave for at least a portion of each of three days amounts to three days. Therefore, Jesus was crucified on Friday.

This is the majority opinion, but some disagree with this. They note that Jesus said that He would be in the earth three days and nights, as Jonah had been in the fish (Matt. 12:40). If taken quite literally, a Friday Crucifixion would not suffice. To resolve this conflict, some have suggested that the Crucifixion was not on a Friday, but instead happened on a Thursday or even a Wednesday. As for the mention that the day following the Crucifixion was a Sabbath, they reason that this actually was the Passover, which was treated as a Sabbath. Support for this view is claimed from John 19:14, where it is recorded that the Crucifixion was on the "day of Preparation, the day before Passover." However, the day of Preparation refers to the day before the Sabbath, not Passover. Furthermore, while a Thursday Crucifixion would have placed Jesus in the tomb for three nights, it would have placed Him in the tomb for four days according to Jewish temporal reckoning.

Many who support the Friday Crucifixion observe that in the first century, people in northern Israel reckoned the beginning and ending of the day differently from how the Jews in Jerusalem and Judea did. Most Jews in Judea, including the Sadducees, followed the ancient Hebrew definition of the day beginning and ending at sundown (a practice continued among Jews today). However, according to ancient sources such as the Mishna, the Galilean Jews and the Pharisees at this time had adopted the convention of beginning and ending the day at sunrise. This resulted in the Galileans and Pharisees observing Passover a day earlier than the Judeans and Sadducees did. Jesus and most of His disciples were from Galilea in the north, so they likely followed the early Passover custom. While the ruling sect among the Jews (the Sadducees) hailed from Jerusalem, they permitted this double observance, because Jerusalem was mobbed by large crowds at the time of Passover, and with so many people wanting to offer their lamb sacrifices, it made the logistics of handling this demand more manageable. This reasoning also resolves other issues. It is clear that the Last Supper was a Passover meal (Matt. 26:17–19; Mark 14:12–16; Luke 22:7–13; John 13:1–4). Jesus instructed His disciples to prepare for the Passover meal, which would have included the sacrificed lamb. The Passover meal began at sundown, and it would have gone on into the evening. But according to the timeline established in all four Gospels, the Crucifixion was on the morning *after* Jesus and His disciples had already prepared and observed the Passover. This perceived contradiction is easily resolved if Jesus and His disciples, being Galileans, observed Passover a day earlier than the day on which the Passover was commemorated by the Judeans.

There has been much scholarly work devoted to the question of which day of the week Jesus was crucified, and it is not our intention to exhaustively discuss it here. For our purposes, we will adopt the majority opinion that Jesus was crucified on Friday. This would have required that Passover was on the Sabbath that year. The leaders of the Jews were concerned with the propriety of leaving dying men on crosses on the Sabbath. This would have been doubly true, because this Sabbath also was the Passover. The leaders of the Jews (who were predominantly Sadducees) probably were unconcerned that they were putting Jesus to death on the Galilean Passover, because the Galilean Jews were often looked upon by the Jewish leaders with some measure of derision.

This sets up a beautiful picture of what transpired at Calvary. As an observant Jew, Jesus had celebrated his final Passover a few hours before He was arrested by the temple guard. Jesus used this meal to teach His disciples many important things, such as showing them what true forgiveness is by washing their feet (John 13:1–20), and His instruction that He is the true vine (John 15:1–17). Yet, whereas Jesus and His disciples had already observed the Passover, many people in Jerusalem were preparing for their observation of Passover while Christ was dying on the Cross. At the same time that these Jewish families were offering up their Passover lambs, the ultimate Passover Lamb was being offered up in sacrifice on their behalf.

Possible Dates of the Crucifixion

We can use astronomical information to determine likely dates for the Crucifixion. As we have seen, the Crucifixion probably was on a Friday. This necessitates that the Passover was on Saturday. (There are a few who believe that the Crucifixion was on Wednesday or Thursday, requiring that Passover was on a Thursday or Friday that year.) From historical and biblical records, we do not know the exact years of Jesus's ministry or of the Crucifixion. However, Bible historians generally give the widest possible range of dates of the Crucifixion as AD 26–AD 36. We can use modern computer programs to determine which day Passover was during each of the years in this range. Obviously we can eliminate any years that Passover fell on Sunday, Monday, Tuesday, or Wednesday. The following table gives the dates of Passover during this range of years. However, note that a lunisolar calendar requires the insertion of an intercalary month roughly every third year. This was done after the twelfth month of one year and before the first month of the New Year. At least since the Middle Ages, the Jews have used the Metonic cycle to determine exactly when to insert an intercalary month. However, it is believed that in biblical times the need for an intercalary month was determined by the condition of the barley crop. Therefore, some of the Passover dates in March on this table may be a month early.

Assuming that these dates are correct, notice that there are only three years within this range on which the first day of Passover was on Saturday, AD 26, 30, and 33. The year AD 26 has few adherents, but the years AD 30 and 33 have many supporters. For those who believe that the Crucifixion was on a Thursday, AD 27 is the only possible year. Similarly, for a Wednesday Crucifixion, the only year was AD 34. The years AD 28, 29, 31, 32, 35, and 36 would appear to be eliminated as possibilities.[1]

---

[1] For further reading on the date of the crucifixion, see Harold W. Hoehner, *Chronological Aspects of the Life of Christ* (Grand Rapids, MI: Academie Books, 1977), as well as Andreas J. Köstenberger and Justin Taylor, *The Final Days of Jesus: The Most Important Week of the Most Important Person Who Ever Lived* (Wheaton, IL: Crossway, 2014).

Darkness on the Land for Three Hours

Mark 15:25 records that the Crucifixion began at the third hour of the day. In the first century the Jews reckoned 12 hours during the day commencing with sunrise, as well as 12 hours at night, beginning with sunset. Therefore, the third hour of the day was 9:00 a.m. in our reckoning. All three synoptic gospels record that there was darkness across the land from the sixth to the ninth hour—noon to 3:00 p.m. (Matt. 27:45; Mark 15:33; Luke 23:44–45). At the end of the darkness, Jesus died, so He was on the cross for six hours. The only time reference with regard to the Crucifixion that the Gospel of John gives is that Pilate pronounced judgment on Jesus at the sixth hour (John 19:14). According to the common Jewish reckoning of time, that would have been noon, but according to the synoptic gospels, Jesus already had been on the cross for three hours at that time. The solution to this dilemma likely is that John here was using the Roman convention of time reckoning. The Roman convention is similar to ours today, in that it started from midnight, making this 6:00 a.m., three hours before the sentence actually was carried out. Since Pilate, a Roman official, was the main character in the first part of John 19, this treatment of time makes sense.

What happened to make the sky dark for three hours during the Crucifixion? There has been much speculation. Of course, those who doubt the authority of Scripture generally believe that nothing of significance happened and attempt to explain away such things. For instance, some observe that there was a partial lunar eclipse on April 3, AD 33, which is a popular choice for the date of the Crucifixion. Some scholars have suggested that early written accounts of the Crucifixion included mention of this lunar eclipse. However, a later scribal error converted the lunar eclipse into a solar eclipse, which then in turn was morphed into the mention of darkness for three hours as contained in the synoptic Gospels. Clearly, this explanation is quite a stretch, and it displays a fundamental lack of respect for the Bible.

Some people have suggested a total solar eclipse to account for the three hours of darkness. However, there are at least two problems with this. First, the maximum length of totality for a solar eclipse is seven minutes, but this is for locations on the equator. At the latitude of Jerusalem, the maximum duration of a total solar eclipse would be closer to five minutes. More importantly, the Crucifixion was at the time of Passover. Passover always is at full moon, but a solar eclipse can happen only at new moon. New moon and full moon are separated by two weeks, so obviously a total solar eclipse could not be the reason for darkness at the time of the Crucifixion.

It appears that there is no known astronomical reason for this darkness. Therefore, some have suggested that the sky was very overcast for three hours. While heavy cloud cover can make the sky dark, its mention by all three synoptic Gospels suggests that this was far darker than any mere cloud cover normally could explain. If this was due to cloud cover, what would have been remarkable about that? Luke 23:45 states that the curtain in the Temple tore when the sun's light failed. There were other apocalyptic signs when Jesus died too. In addition to the torn Temple curtain, Matthew 27:51–53 says that the earth shook, rocks split, and many saints came back to life. These are unusual, inexplicable, and perplexing. Extreme and unusual darkness often accompanies apocalyptic signs (cf. Isa. 5:30; 13:10–11; Joel 2:1–2; Amos 5:20; Zeph. 1:14–15; Matt. 8:12; 22:13; 25:30). Hence, while clouds may have played a role, this darkness was far greater than clouds normally could account for.

While it is tempting to attempt to explain this darkness by some sort of physical mechanism, and hence a mechanism that we can understand, that sort of explanation robs the darkness of what it signified. At least some of the other signs at the time of Jesus' death certainly do not appear attributable to natural phenomena. There may be a parallel between the three hours of darkness at the Crucifixion and the three days of darkness during the ninth plague in Egypt (Exod. 10:21–29). That darkness was so severe as to be felt. The Egyptians could not even see one another. Even during the darkest night, people can see by starlight. During the ninth plague, darkness extended day and night for three days. Throughout the ninth plague, the Israelites had light where they lived. No cloud cover or sand storm (as some have suggested) could possibly explain the ninth plague. Hence, the ninth plague probably was a supernatural event. Just as the ninth plague in Egypt was supernatural, the three hours of darkness during the Crucifixion probably was supernatural.

An Alleged Lunar Eclipse the Night Jesus Died

Historians generally date the Crucifixion within the range of AD 26–AD 36. This comes from considerations of the biblical clues about the reigns of Herod, Pontius Pilate, Tiberius, and Quirinius. In the estimation of some people, the late date of Herod's death is important to the thesis of the April 3, AD 33 Crucifixion. For a long time historians have thought that Herod died in 4 BC, and they generally continue to think so. If Herod died this early, then Jesus had to be born in or prior to 4 BC, even as early as 6 BC. Luke 3:23 tells us that Jesus started his ministry when he was about 30 years old, and the chronology of the Gospels suggest that His ministry lasted (at minimum) a little more than three years. Assuming a birth year of no later than 4 BC and recognizing that there was no year zero (1 BC was immediately followed by AD 1), we arrive at a Crucifixion date not much later than AD 30. However, this is problematic, because Luke 3:1–3 records that John began his ministry in the 15th year of the reign of Tiberius. Tiberius became emperor in the year AD 14, so it appears that John began his ministry around AD 28. John obviously baptized Jesus after he had begun his own ministry but before Jesus began His ministry. This scarcely would leave two years for Jesus' ministry, if the Crucifixion were in AD 30. On the other hand, Luke did not state that Jesus was *exactly* 30 years old when He began His ministry, but that Jesus was "about 30 years of age." This is consistent with an age of 33 years, so suppose that was Jesus' age when He began His ministry. If Jesus was born in 5 BC, then He would have begun His ministry near the year AD 29. If Jesus' ministry lasted a little more than three years, then AD 33 is the proper date for the Crucifixion.

On the other hand, some have argued that Herod died later, in 1 BC, which supposedly gives a better match to the AD 33 Crucifixion date; but as we have seen, this is not necessary. The date of Herod's death is computed from a mention in Josephus that Herod died shortly before Passover, and that his death was preceded by a lunar eclipse. There were only two lunar eclipses shortly before Passover visible from Israel during the possible range of Herod's death, one in 4 BC and the other in 1 BC. Most historians continue to believe the earlier date is correct, but in recent years some Christians have endorsed the later date of Herod's death. Much of this motivation appears to come from a particular theory about the Christmas star, a topic discussed in Chapter 7. There are two arguments put forth in support of the later date of Herod's death. One argument is that the earlier date allows just a month for several events recorded by Josephus to occur (such as the execution of Matthias and Judas and their followers the day preceding the eclipse and Herod's illness that led to his death). They argue that there was not enough time to accomplish these events in a month or less, but that the three months of the 1 BC window was enough time. The second argument is that the 4 BC lunar eclipse was partial but that the 1 BC eclipse was total and hence better conforms to what Josephus recorded. However, Josephus did not claim that the eclipse was total, but rather that the moon was eclipsed, so either a partial or total lunar eclipse would fit the description.

In addition to a particular theory about the Christmas star, there is another reason for belief in the 1 BC death of Herod that has gained traction. In recent years there has been renewed interest in the April 3, AD 33 Crucifixion date, because there was a lunar eclipse that evening. As we have already seen, the early date of Herod's death and the AD 33 Crucifixion date are compatible, but a late date for Herod's death would eliminate the only other possible Crucifixion date in AD 30. This would clearly establish April 3, AD 33 as the date of the Crucifixion, and some people see prophetic fulfillment in the lunar eclipse that evening. The most common color of the moon during a total lunar eclipse is some shade of red or orange. Blood also is red, so many people who prefer the AD 33 date view this eclipse as fulfillment of Joel's prophecy of the moon being turned to blood, as quoted later by the Apostle Peter at Pentecost (Acts 2:15–21).

This and other prophetic passages are discussed in Chapter 9, but several things are worthy of note here. There is doubt that Peter's quotation of Joel at Pentecost or even Joel's prophecy itself necessarily refers to a total lunar eclipse. To stand out as a sign (cf. Gen. 1:14), the moon turning to blood must be remarkable. However, a lunar eclipse by itself, though interesting, is hardly remarkable. Furthermore, the context of Peter's quotation of Joel's prophecy was to explain the miracle occurring on Pentecost of how people from various parts of the world were able to understand in their own languages what the Jewish Galileans were saying (Acts 2:4–13). This was preceded by the coming of the Holy Spirit (Acts 2:1–4). Peter related those two events to what Joel had written concerning the outpouring of the Holy Spirit resulting in people prophesying (Joel 2:28–29; Acts 2:17–18). Peter went on to quote the rest of Joel's prophecy (Joel 2:30–32; Acts 2:19–21) about wonders in heaven and earth, proclaiming that salvation shall be made available to people outside of Israel. But the way of salvation for the Gentiles came later, in Acts 10. Apart from Peter's quotation of Joel, there is no mention in Acts of the wonders in heaven and earth that Joel's prophecy foretold. The language from Joel about a darkened sun, a blood-like moon, and upheaval in the earth better conforms to the words of Revelation 6:12, which describes events that many view as still future. Thus there is considerable doubt that Joel's prophecy was *completely* fulfilled at the time of the Crucifixion or at Pentecost

Why does a total lunar eclipse appear red? Because the earth has an atmosphere, its shadow (called the umbra) is not totally dark. Rather, the atmosphere refracts light into the earth's shadow. This is why the sky doesn't immediately turn dark after sunset. Additionally, the earth's atmosphere scatters light that is bent into the umbra. This scattering is dependent upon wavelength, with shorter wavelengths being affected the most. Since in the visible part of the spectrum shorter wavelengths are blue and longer wavelengths are red, most of the blue light is scattered, leaving red to dominate the refracted light. This is why sunset and sunrise normally are red. Scattering also causes the earth's umbra generally to be red—but not always.[2]

The color of total lunar eclipses can vary tremendously from eclipse to eclipse. This variation is attributed to atmospheric conditions along the edge of the earth through which the light travels. The most commonly reported color of a totally eclipsed moon is some shade of red or orange, though yellow is occasionally reported, and some eclipses are virtually black. Most unusual was the eclipse of July 6, 1982, one of the longest total lunar eclipses in history. A portion of the earth's umbra was black as coal, but the rest was peach colored. The earth's umbra normally has uniform color and darkness, so this was a rare eclipse. It is important to note that with the wide variation in color and intensity in total lunar eclipses, it is presumptuous to assume automatically that any particular past or future total lunar eclipse did or will have a blood red appearance.

There are several problems with claiming that the April 3, AD 33 lunar eclipse was a fulfillment of prophecy. First, one must assume that the sky was clear enough over Jerusalem that people could have seen the eclipse. If extreme cloud cover was responsible for the great darkness between noon and 3:00 p.m. the day of the Crucifixion, the sky would have had to have cleared tremendously within three hours, for the moon would not have been visible otherwise.

More importantly, the April 3, AD 33 lunar eclipse was a partial eclipse, not total. This partial eclipse was far from total, and the maximum coverage of the moon occurred half an hour before the moon rose over Jerusalem that evening. When the moon rose, the lower right portion was outside of the earth's umbra, and the partial eclipse ended about 45 minutes after rising. A small portion (far less than half) of the moon to the upper left was in the earth's umbra. A large portion of the moon not in the umbra was in the earth's penumbra. The penumbra is the partial shadow of the earth, where the sun's light is not completely blocked. While the penumbra is not fully illuminated, there is

---

[2] For a diagram illustrating this phenomena, see https://answersingenesis.org/astronomy/moon/will-lunar-eclipses-cause-four-blood-moons-in-2014-and-2015/.

considerable light within the penumbra, thousands of times more than there is in the umbra. It is very difficult for the human eye to notice penumbral shading. This is because the eye adeptly compensates for the relatively small drop in intensity in the penumbra. And there is only the subtlest of color change. Only right along the penumbra's border with the umbra is any difference noted. The very slight color change in the penumbra as compared to the uneclipsed moon would require a very sensitive device for measurement. It is difficult for the human eye to judge the umbra's color, except when the eclipse is nearly total. However, when the moon rose that night in Jerusalem, the eclipse was far from total. In short, in no way could the penumbra be described as red, let alone blood red.

In disputing with the author on this point, some supporters of the prophetic nature of the April 3, AD 33 lunar eclipse have offered photographs of the partially eclipsed moon. In these photographs, the moon appears red. However, in each of the photographs the moon is extremely low in the sky, nearly on the horizon. The redness of the moon in these photographs is due to atmospheric extinction. Atmospheric extinction is the scattering of light as previously discussed. This scattering is much greater at shorter wavelength (blue) light than at longer wavelength (red) light, so this phenomenon systematically makes objects appear redder than they really are. Astronomers call this effect reddening, and they must correct for this in their observations. Objects viewed near the horizon are passing through a tremendous amount of atmosphere, so their light has undergone great scattering and hence great reddening. This is why the rising or setting sun appears red. The rising or setting moon appears red too. To be blunt, those who submit such photographs to prove their point that a penumbral or partially eclipsed moon appears red don't know what they are talking about. If they had spent much time actually observing the moon, they would understand that the moon usually appears red when low in the sky.

If the moon were visible rising over Jerusalem on the evening of April 3, AD 33, would it have appeared red? Probably, but not because of the eclipse. If the moon were not eclipsed that night, it would have looked just as red from atmospheric extinction. The moon rises or sets nearly every evening (only a few nights per month near the new phase is the moon not visible). The rising or setting moon often appears deep red, though how reddened the moon is depends upon the local atmospheric conditions. Since this happens nearly every night, it is unlikely that normal reddening due to the atmosphere would constitute a fulfillment of prophecy. Many people already recognize that the darkness for three hours that afternoon likely was the result of unusual atmospheric conditions. If so, it is very probable that those conditions did not entirely clear by nightfall, which could account for an unusually red moon that evening, even without an eclipse. So this partial eclipse would have had little to do with the redness that might have colored the moon that night.

On the other hand, proponents of this view claim that the partial eclipse of April 3, AD 33 accomplished something that only a total eclipse could do: give the moon a red appearance. However, it is inconsistent for supporters who claim the AD 33 partial lunar eclipse had prophetic significance to also claim that the partial eclipse that night could accomplish what it could not while simultaneously maintaining that the partial eclipse of 4 BC was not significant enough to have warranted mention by Josephus.

Again, those who propose that the partial lunar eclipse on the evening of April 3, AD 33 was the fulfillment of Joel's prophecy mean well, but there is a downside. Many Christians find significance and meaning in that astronomy may offer some evidence of events surrounding the Crucifixion. However, lunar eclipses are not that rare and a partial eclipse fails to qualify as a blood moon. If a lunar eclipse is what Peter was referring to when he quoted the prophet Joel at Pentecost, then that hardly was a remarkable fulfillment of prophecy. The skeptic can just as easily respond that early Christian leaders merely wove current natural events (such as a lunar eclipse) into their narrative, and thus this proves nothing. Either approach conforms to a particular worldview. Ultimately, the appeal to a partial lunar eclipse fails to explain the moon being turned to blood, so Christians ought to avoid this argument.

CHAPTER 9

Astronomical Aspects of Prophetic Literature and the End Times

Different Views of Biblical Eschatology

According to Genesis 1:14, one of the purposes of the heavenly bodies is to be for signs. We often understand these signs in apocalyptic terms. If so, then it is inevitable that interpreting signs in the heavens will be within one's understanding of biblical eschatology. Eschatology is the study of future events within the context of Scripture. There are many different views of eschatology. It is not the purpose of this book to endorse one particular position. Nor is it this book's purpose to discuss fully the various positions. On the other hand, it would be difficult to discuss signs in the heavens without at least defining some of the prominent ideas with regard to eschatology, so we shall briefly outline some positions here. One should not construe anything in this brief introduction as endorsing any particular position, nor should this discussion be considered exhaustive.

Many Christians believe that in the future there will be a literal thousand-year reign of Christ on this earth, followed by an eternal state with a New Jerusalem on a New Earth. This literal thousand-year reign commonly is called the *Millennium*, from the Latin word for one thousand. The major passage used in defense of this position is Revelation 20. Revelation 20:1–3 speaks of the binding of Satan for a thousand years. Revelation 20:4–6 tells of resurrected saints who were martyred for Jesus and the word of God reigning with Jesus on earth during these thousand years. Revelation 20:7–10 says that after the thousand years, Satan will be loosed to deceive the nations once more, but that he will be cast into the Lake of Fire upon his defeat. Revelation 20:11–14 tells of the Great White Throne judgment. Revelation 21:1–8 goes on to describe the New Heaven and the New Earth, while Revelation 21:9–22:5 tells specifically of the New Jerusalem.

Closely related to the belief in a literal thousand-year reign of Christ on the earth is a belief that it will be preceded by the *Tribulation*, a seven-year time period which will see the rise of a one-world dictator called the Antichrist. Those who believe in a literal thousand-year reign of Christ commonly hold that many of the events described in Revelation 4–19 will occur during the Tribulation. Support for this position is claimed from Old Testament passages as well. For instance, the Tribulation is identified with the 70th week of Daniel's prophecy (Dan. 9:24–27), linking it to time periods mentioned in the book of Revelation (Rev. 11:2–3; 12:14; 13:5). Furthermore, certain prophetic passages from the Old Testament that do not appear yet to have had a literal fulfillment are relegated to future fulfillment during the Millennium. For instance, Isaiah 11:6–12 describes an idyllic time when carnivorous animals will peaceably abide with what had formerly been their prey, and when venomous snakes will pose no danger. Isaiah 2:2–7 and Micah 4:1–8 say that in the latter days the Lord will settle disputes for the nations and that war will be no more. This, too, is supposed to be during the Millennium.

Another event closely related to the Tribulation is the *Rapture*, the time when Jesus returns to the earth (Acts 1:11), and the saints, both alive and dead, will be caught up into heaven. Though the word *rapture* does not appear in the Bible, this term, meaning "to snatch away," refers to the event described by the Apostle Paul in 1 Thessalonians 4:13–18. The language of 1 Corinthians 15:50–52 is similar enough to suggest that it describes the Rapture as well. (Some people think that Matt. 24:36–44 also refers to the Rapture, though many more think that this refers to the Lord taking people away in judgment.) Exactly when the Lord will return and the when Rapture will occur is much debated. Many who believe in a literal millennial reign believe that this will be before the Millennium. Hence, belief in a Rapture before a literal millennial reign of Christ is called *premillennialism*. Even among premillennialists, there is much debate about the time of the Rapture in relation to the Tribulation. Some think that the Rapture will precede and thus usher in the Tribulation. Others think that the Rapture will occur at the end of the Tribulation. Still others think that the Rapture will happen sometime during the Tribulation. These three positions are called, respectively, *pretribulational, posttribulational,* and *midtribulational* (though some midtribulational proponents prefer the term *pre-wrath tribulation*). Most posttribulational premillennialists hold to what is called "historic premillennialism," which depends on Covenant Theology and the replacement of Israel with the church. Pretribulation and midtribulation views are almost always associated with Dispensational Theology. The reason for this is because, as it is argued, God's attention will revert to Israel once the church is removed from the earth at the time of the Rapture.

However, not all Christians agree with the premillennial view. The *postmillennial* view holds that Christ will return to earth after His thousand-year reign, but that this reign will be spiritual rather than physical (cf. Luke 17:20–21; John 18:36). This reign may or may not be a literal 1,000 years. Postmillennialists believe that by the church fulfilling the Great Commission most people eventually will be saved. This will bring about 1,000 years of this spiritual reign, in which a Christian ethic will govern the world, resulting in peace and prosperity. At the conclusion of this thousand years, Christ will return to earth, upon which there will be the resurrection of the dead, the judgment, and the eternal state. Postmillennialism was popular in the past, but it has waned considerably among conservative Christians over the past century and a half. Pivotal in the decline of postmillennialism was the carnage of the First World War and the aftermath of that war.

Amillennialism is in some aspects similar to, yet distinct from, postmillennialism. Like some postmillennialists, amillennialists reject a literal thousand-year reign of Christ on earth. They believe instead that the millennial reign of Christ is within the hearts of believers in the church age. However, amillennialists are more pessimistic about the outward success of that reign. Unlike postmillennialists, who believe that improvements in the world brought about by this reign will bring about Christ's return to earth, amillennialists believe that at some point Christ will return, but not because of the complete success of the church in obeying the Great Commission. Amillennialists believe that good and evil will remain mixed within the church age. The parable of the wheat and tares (Matt. 13:24–30) is taken as support of this position. In this regard, many amillennialists agree more with premillennialists in that there will be a falling away (cf. 2 Thess. 2:3) toward the end of the church age.

While postmillennialists disagree with premillennialists as to how literal the 1,000 year reign of Christ on earth will be, they can agree (though not necessarily) with premillennialists that His reign will be exactly 1,000 years. However, amillennialists cannot accept that the non-literal reign of Christ will be exactly 1,000 years, because in their estimation the reign of Christ began in the first century AD and nearly two millennia have transpired since. Their answer to this difficulty is that the 1,000 year duration need not be literal but merely indicative of some long period of time. As for the binding of Satan in Revelation 20, amillennialists believe that Satan has been bound during the church age in that the church has grown during this time. That is, Satan has not been able to destroy the church because of God's restraint. In this respect, Christians have remained more true to God than Israel managed to, because Satan was not restrained in the Old Testament, nor did believers then have the advantage of the indwelling of the Holy Spirit as the church does today.

How do these views relate to the interpretation of biblical prophecy?

Premillennialists tend to view the fulfillment of end-times prophetic passages very literally, while postmillennialists and amillennialists generally do not. As already discussed, premillennialists generally think that much of the book of Revelation has yet to be fulfilled. Some amillennialists and many postmillenialists think that much of the book of Revelation was fulfilled in the first century. A popular view with supporters of these positions is that many of the things that premillennialists believe will happen during the Tribulation were fulfilled during the siege of Jerusalem and the resulting destruction of the Temple in AD 70. Thus, there is disagreement as to when John wrote the book of Revelation. Premillennialists generally hold that John wrote Revelation around AD 95, more than two decades after the Temple's destruction. However, some amillennialists and postmillennialists maintain that John wrote it about three decades earlier, prior to the Temple's destruction.

Everyone agrees that prophetic passages often employ imagery and symbolic language. However, premillennialists tend to think that eschatological prophecies will be fulfilled literally, while postmillennialists and amillennialists tend to think that many prophecies anticipate a figurative fulfillment. This is true not only of the book of Revelation, but of many Old Testament passages as well. For instance, some Old Testament passages concerning the Temple do not appear to have been literally fulfilled (e.g., Isa. 2:2–4; Ezek. 43:1–7; Hag. 2:9). Therefore, if these are to be literally fulfilled, then that fulfillment must be in the future. Premillennialists place this fulfillment during the Millennium. On the other hand, if the fulfillment is not literal, then many, if not all, of these passages could have been fulfilled already. Some amillennialists and postmillennialists think that many of these prophecies were or are fulfilled symbolically in the universal church.

Another view is that some of the Old Testament prophecies concerning Israel which do not appear to have been fulfilled never will be fulfilled, because those prophecies were part of the Old Testament covenant with Israel. That covenant, it is argued, always was dependent upon Israel's obedience. Since Israel in its disobedience has rejected their covenant, in this view God has replaced that covenant with the New Covenant. Since the covenant with Israel has been effectively withdrawn, the prophecies will not be fulfilled.[1] This leads to replacement theology, the belief that the church has replaced Israel. In replacement theology, there is no future for national Israel in God's plan. Dispensational premillennialists reject replacement theology, believing that God's promises to Israel were unconditional. Israel is, in this view, the recipient of the New Covenant, and the church participates in the blessings of (but does not receive) that covenant. For the present time, God's focus is on the church. However, there will be a future time when God again will deal with Israel (cf. Rom. 11:11–24). Dispensational premillennialists believe that with the church's removal at the Rapture, God once again will work through Israel throughout the Tribulation and the Millennium.

With this very brief introduction to some various schools of thought concerning eschatology, let us now turn to biblical passages about signs in astronomical bodies. With such a wide diversity of viewpoints on the prophetic passages that concern the end times, there is much difference of opinion concerning what these passages mean.

Apocalyptic Signs in the Sky

Many prophetic passages include references to signs in the sky, consistent with one of the purposes of the heavenly bodies given in Genesis 1:14. For instance, in Luke 21:25–28, Jesus said,

> "And there will be signs in the sun and moon and stars, and on the earth distress of nations in perplexity because of the roaring of the sea and the waves, people fainting with fear and with foreboding of what is coming on the world. For the powers of the heavens will be shaken. And then they will see the Son of Man coming in a cloud with power and great glory. Now when these things begin to take place, straighten up and raise your heads, because your redemption is drawing near."

---

[1] Those who disagree with this view say that it conflates the unconditional Abrahamic Covenant (with the land grant, the perpetual protection clause, and the blessing promise) with the conditional Mosaic or Sinaitic Covenant.

The ascension account of Acts 1:9–11 reveals that Jesus will return to earth in a similar manner to how He left (in a cloud). Therefore, it appears that Jesus' statement about His coming in power and glory recorded in Luke is the same event that the angel in the book of Acts was talking about. Since Jesus physically left the earth, His return must be physical as well, thus eliminating the possibility of a mere spiritual return, as some have suggested.[2] While Christians of different eschatological persuasions may disagree on whether various prophetic passages have yet been fulfilled, the fact remains that Jesus has not yet returned to earth in a cloud as He left this earth, so this must be in the future. What might these signs in the sun, moon, and stars be that Jesus associated with His return?

Matthew 24:29–31 is a parallel passage to Luke 21:25–28, and it offers more detail. It reads,

> "Immediately after the tribulation of those days, the sun will be darkened, and the moon will not give its light, and the stars will fall from heaven, and the powers of the heavens will be shaken. Then will appear in heaven the sign of the Son of Man, and then all the tribes of the earth will mourn, and they will see the Son of Man coming on the clouds of heaven with power and great glory. And he will send out his angels with a loud trumpet call, and they will gather his elect from the four winds, from one end of heaven to the other."

Notice that this passage adds three details not found in the Luke 21 passage. First, it gives more detail with regard to the signs in the sun and moon—they will be darkened. Second, it elaborates on the activity of the stars, noting that they will fall from heaven. Third, it states that Jesus will send His angels with a loud trumpet call to gather His elect.[3] Another parallel passage to these two is Mark 13:24–27, but it does not present any additional information aside from what is offered in the other two synoptic Gospels.

We must be careful in handling these parallel passages and the verses surrounding them. Some Christians lump together all of Jesus' words here (Matt. 24:4–51; Mark 13:3–37; Luke 21:8–36),[4] but Jesus responded to more than one question that his disciples posed. What is the context? In response

---

[2] For instance, the Jehovah's Witnesses cult teaches this.

[3] The trumpet call and gathering of the saints lead some to conclude that this is the same event mentioned in 1 Thessalonians 4:13–18.

[4] Since the words of Jesus Christ here were delivered on the Mount of Olives, they are known as the Olivet Discourse.

to the disciples' pointing out the elegance of the Temple, Jesus commented upon the fact that the Temple would be destroyed (Matt. 24:1–2; Mark 13:1–2; Luke 21:5–6). Matthew 24:3 records that afterward the disciples came to Jesus privately on the Mount of Olives and asked, "Tell us, when will these things be, and what will be the sign of your coming and of the end of the age?" Mark's Gospel concurs on the location (Mount of Olives), but it states that it was Peter, James, John, and Andrew who asked, "Tell us, when will these things be, and what will be the sign when all these things are about to be accomplished?"[5] Finally, Luke 21:7 does not identify the location where the questions were asked, but merely records that the disciples asked Jesus, "Teacher, when will these things be, and what will be the sign when these things are about to take place?" Thus the three synoptic Gospels record that the disciples asked two questions about the things that Jesus had just commented on (the destruction of the Temple): (1) what time the events would happen, and (2) what the accompanying signs were to be. However, the account of Matthew has the greatest detail with regard to the questions (and the answers to the questions).

Today we easily can separate the two questions, because we understand that Jesus would die and rise again within days of saying these words, that He would leave the earth less than two months later, that the Temple would be leveled a few decades later in AD 70, and that Jesus has not yet returned nearly two millennia later. However, the disciples at this time knew none of this. They probably could not fathom that Jesus was going to die in a matter of days or that these other events would happen. Indeed, they seemed to be troubled by Jesus' statement that the Temple would be destroyed, which prompted their questions. In their limited view of eschatology, the disciples probably envisioned Jesus fulfilling all the prophecies very soon (Luke 19:11 indicates that they supposed that the Kingdom of God was to appear immediately). Since in their minds this would have involved the Temple, they could not fathom the coming destruction of the Temple. To say that they were confused would be putting it mildly. Perhaps they thought that they were asking one question, thinking that the two events were synonymous. However, we now know that the Temple's destruction and the Lord's return are not the same, at the very least because they are separated in time by nearly two millennia. Indeed, the disciples did

---

[5] Three of these four disciples, Peter, James, and John, frequently are mentioned in the Gospels and amounted to a sort of inner circle among Jesus' disciples. As such, they probably posed the question on behalf of all of the disciples. Andrew's inclusion in asking Jesus for clarification is an anomaly.

not anticipate Jesus' ascension, so when they asked about Jesus' coming, they probably did not intend His Second Coming, though we understand it that way today. Instead, by asking about Jesus' coming, the disciples probably had in mind Jesus' assumption of power.

We may be confused too, so perhaps we ought not to be so quick to fail to appreciate the disciples' dilemma. For instance, some people today see in Matthew 24:3 three questions instead of two: the timing and signs of the Temple's destruction, the signs of Jesus' return to earth, and the signs of the end of the age. This assumes that the Lord's return will not usher in the end of the age, for the end of the age could be sometime after the Lord's return, such as in a pretribulational eschatology. However, many others disagree with this, seeing two questions rather than three. If that were not confusing enough, there is much disagreement with regard to which parts of Jesus' reply apply to the first versus the second (or even the third) question. If nothing else, this disagreement today and the lessons of history ought to cause us to be more humble in our attitude toward eschatology, less dogmatic in our conclusions, and more understanding of those, such as the disciples, who we think may not see as well as we do.[6]

Interestingly, early in the Olivet Discourse, Luke 21:11 says,

"There will be great earthquakes, and in various places famines and pestilences. And there will be terrors and great signs from heaven."

The parallel statements in Matthew 24:7 and Mark 13:8 omit the phrase "and great signs from heaven." This phrase fits in well with Luke 21:25. Does this mean that Luke 21:11 and Luke 21:25 refer to the same things, or does this passage refer to two separate incidences of signs from heaven?

Several Old Testament passages may refer to the same cosmic events as those mentioned in the Olivet Discourse. For instance, Isaiah 13:9–10 says,

Behold, the day of the LORD comes,
    cruel, with wrath and fierce anger
to make the land a desolation
    and to destroy its sinners from it.
For the stars of the heavens and their constellations
    will not give their light;
the sun will be dark at its rising,
    and the moon will not shed its light.

---

[6] Virtually no one in the first century anticipated the details of the First Advent of Christ. They had expected something very different from what happened. Is it not presumptuous and arrogant to think that we in the 21st century have much clearer vision with regard to the Second Advent?

But there are some differences. The moon is darkened as in the Olivet Discourse, but the sun is not mentioned in these verses. Furthermore, the stars are said to be darkened rather than falling from heaven as in Matthew 24:29–31.

Or consider Isaiah 24:21–23:

> On that day the LORD will punish
>   the host of heaven, in heaven,
>   and the kings of the earth, on the earth.
> They will be gathered together
>   as prisoners in a pit;
> they will be shut up in a prison,
>   and after many days they will be punished.
> Then the moon will be confounded
>   and the sun ashamed,
> for the LORD of hosts reigns
>   on Mount Zion and in Jerusalem,
> and his glory will be before his elders.

The effect on heavenly bodies here is vague—it only states that the host of heaven, generally understood to be the stars, will be punished, that the moon will be confounded, and that the sun will be ashamed. Given that one of the purposes of the heavenly bodies is to provide light on the earth (Gen. 1:15), it is easy to see that the punishment of stars, the confounding of the moon, and the shaming of the sun could refer to their inability to perform this function due to darkening. So this could refer to the same events of the Olivet Discourse. However, the description is a bit different, so perhaps this refers to something else. Either way, it appears to describe something ominous, quite out of the ordinary operation of things in the heavens.

A much better match appears in Joel 2:10:

> The earth quakes before them;
>   the heavens tremble.
> The sun and the moon are darkened,
>   and the stars withdraw their shining.

In Joel 1, the prophet describes an invasion of locusts as God's judgment on Israel. In Joel 2:1–11 the prophet uses the locust invasion to transition to a discussion of "the day of the LORD" (Joel 2:1), demonstrating that the day of the LORD would be more ominous than the locust invasion. That day is described as "a day of darkness and gloom, a day of clouds and thick darkness" (Joel 2:2).[7] Swarms of locusts can briefly darken the sky, but the darkness accompanying the day of the LORD seems more threatening than that, and thus can be explained by the sun, moon, and stars withdrawing their shining, as stated in verse 10. Joel 3:15 returns to this theme of the darkening of the sun, moon, and stars in the day of the LORD by repeating the second part of Joel 2:10 word for word. The parallel to Matthew 24:29–31 is so strong as to suggest that they refer to the same event(s).

Amos 8:9 similarly says,

> "And on that day," declares the Lord GOD,
> "I will make the sun go down at noon
> and darken the earth in broad daylight."

Some people may insist that this verse means that the sun literally will set at noon. However, it need not be. Even today, in referring to a total solar eclipse, people often describe it as midnight at noon, or the sun setting at noon. That is, we sometimes use figurative language to describe what literally happens. Since prophetic passages are filled with figurative language of this type, Amos 8:9 may be employing figurative language to describe a frightful thing. Zephaniah 1:15–16 likewise states,

> A day of wrath is that day,
> a day of distress and anguish,
> a day of ruin and devastation,
> a day of darkness and gloom,
> a day of clouds and thick darkness,
> a day of trumpet blast and battle cry
> against the fortified cities
> and against the lofty battlements.

Ezekiel 32:7–8 is similar to the passages that speak about the darkening of the sun, moon, and stars:

> "When I blot you out, I will cover the heavens
> and make their stars dark;
> I will cover the sun with a cloud,
> and the moon shall not give its light.
> All the bright lights of heaven
> will I make dark over you,
> and put darkness on your land,
> declares the Lord GOD."

---

[7] A similar description appears in Amos 5:18–20.

If one were to read these two verses in isolation from their context, it would be easy to equate them with some of the other passages just discussed. However, it is important to look at the context of this passage (or any other verse, for that matter). According to the introduction in verse 2, this chapter is about the judgment of Pharaoh, king of Egypt. In the larger context, Ezekiel 32 is the last of four chapters (29–32) concerned with the judgment of Egypt. Ezekiel 30:10 names Nebuchadnezzar as the instrument of God's judgment, and the king of Babylon is mentioned again as executioner in chapter 32 (verse 11). Chapter 31 concludes with a pronouncement of death on Pharaoh, leading to the detailed lament of Pharaoh in chapter 32. Nebuchadnezzar lived in the late seventh and early sixth centuries BC. Cyrus the Great conquered Babylon in 539 BC, after which Babylon's influence greatly waned. A few decades after Alexander the Great conquered that part of the world in the late fourth century BC, Babylon ceased to exist. For that matter, no Pharaoh has ruled Egypt in nearly as long. Therefore, it would seem that this judgment was carried out in the sixth century BC, long before Jesus' earthly ministry. If so, then the darkness of Ezekiel 32:7–8 cannot so easily be equated with the darkness that Jesus spoke of in Luke 21:25–28.[8]

Notice that Ezekiel 32:7 begins with "When I blot you out…." This is only one of 18 instances where the pronoun *you* or *yours* appears in verses 2–8. Who is being addressed here? Verse 2 makes it very clear that it is Pharaoh. Verse 8 says that God will "make dark over you, and put darkness on your land." Since the word *land* is modified by the possessive pronoun *your*, this must refer to Pharaoh's land, Egypt. The significance of darkness is easy to miss. The sun god *Ra* was one of the most important Egyptian deities. The Pharaohs were considered to be representatives of *Ra*. In this sense, Pharaoh was a source of light in ancient Egypt, and so the death of Pharaoh represented loss of light in the kingdom, albeit in a metaphorical sense. These considerations help put into context the ninth plague in Egypt, the plague of darkness (Exod. 10:21–29). It was the ultimate insult to Egypt when God struck Egypt with darkness, because it underscored the impotence of Egypt's god-king before the true and living God. The ten plagues increased in intensity, culminating in the tenth plague. It is not coincidental that the ninth plague was followed by the tenth

---

[8] However, one might argue that some prophecies have dual fulfillment—an immediate partial fulfillment followed by a later complete fulfillment—and that Ezekiel 32:7–8 may refer to same events as those in the Olivet Discourse.

and final plague, the death of the firstborn in each Egyptian household.[9] Thus, Ezekiel 32:7–8 is a polemic against Egypt and its Pharaoh. Thus this particular mention of darkness in heavenly bodies in Ezekiel 32:7 likely is not literal but figurative.

This underscores the care that one must take in interpreting signs in the heavens. One must carefully consider the context, as here in Ezekiel 32:7, before concluding that similar language appearing in two or more passages implies coincidental prophetic events. On the other hand, prophetic passages need not be exactly the same in order for them to describe the same future events. Prophecy frequently employs imagery, and so the imagery of one passage may differ from that of another, even when both are describing the same prophetic event(s). Finally, some prophetic passages may be figurative rather than literal, as in Ezekiel 32:7.

Are the events of most of these passages, the darkening of heavenly bodies, to be taken literally? Notice that the darkening of heavenly bodies is accompanied by terrifying and perplexing events, events that appear to be real, not imagery. In that context, figurative, non-literal darkening would not suffice. Is the prophesied darkening due to something that happens in the heavenly bodies or to the atmosphere of the earth to block their light? This is not so clear. Both Joel 2 and Amos 5 comment on the dark and gloomy nature of the day, and clouds are mentioned as well. Furthermore, the context of Joel 2 suggests the sky being darkened by clouds of locusts. This suggests that this darkening is atmospheric, but with marked differences from normal cloudy days. Perhaps the darkness is not only profound, but also widespread and persistent. Some people have suggested that this will be the result of multiple, large volcanic eruptions, or multiple nuclear explosions, or impacts from asteroids that will kick up large amounts of dust into the sky. Normally, dark cloud cover does not last for a long time, but instead is a temporary condition. Nor is it over a large geographical area. At any rate, this is not some normal cloud cover.

All three synoptic Gospels record extreme darkness at the time of the Crucifixion (Matt. 27:45; Mark 15:38; Luke 23:44). There is no specific prophecy for which this would have been a fulfillment, though some have suggested that this darkness was at least a partial fulfillment of Amos 8:9. The mention of this darkness implies that it was profound and remarkable. Furthermore, Matthew 27:51–54 includes mention of other miraculous events, such as the tearing of

---

[9] For more information on this point, see fn. 7 in Chapter 1.

the Temple veil, an earthquake, and the resurrection of dead saints. However, the accounts of Mark and Luke include only the tearing of the veil (Mark 15:38; Luke 23:45).

Besides the aforementioned passages, there are other passages that mention darkness as a sign of God's judgment. An example is Isaiah 5:30. The immediate context (Isa. 5:26–30) is that of foreign nations being the instrument of God's judgment upon His people. This principally was fulfilled by Assyria and Babylon. Other examples of darkness as part of God's judgment are Isaiah 8:22, Matthew 8:12, and Matthew 22:13. Whatever the cause of this prophesied darkness, it will be remarkable and perplexing, or else it hardly would constitute a sign.

Not surprisingly, the book of Revelation mentions the darkening of heavenly bodies too. Revelation 8:12 reads,

> The fourth angel blew his trumpet, and a third of the sun was struck, and a third of the moon, and a third of the stars, so that a third of their light might be darkened, and third of the day might be kept from shining, and likewise a third of the night.

This appears to refer to a dimming of the sun, moon, and stars, which matches well with cosmic signs of the Olivet Discourse, as well as some of the aforementioned Old Testament passages. Hence, some have concluded that this passage describes the same events of those other verses. (Interestingly, later in the book of Revelation [16:8–9], there is the implication that the sun's brightness will increase.)

The book of Revelation specifically mentions "a sign in heaven" three times. The first is in Revelation 12:1–2:

> And a great sign appeared in heaven: a woman clothed with the sun, with the moon under her feet, and on her head a crown of twelve stars. She was pregnant and was crying out in birth pains and the agony of giving birth.

This sign in heaven is immediately followed by the second sign in heaven mentioned in the book of Revelation. Revelation 12:3 reads,

> And another sign appeared in heaven: behold, a great red dragon, with seven heads and ten horns, and on his heads seven diadems.

The remainder of chapter 12 describes the interaction of these two signs in heaven. The dragon is identified as Satan in verse 9, but the woman and her child are not identified. Many people see the woman as Israel and her child as Jesus (cf. Gen. 37:9). A minority viewpoint is that the woman in the beginning of Revelation 12 is Mary, the mother of Jesus.

The final sign in the sky in the book of Revelation is in Revelation 15:1, which reads,

> Then I saw another sign in heaven, great and amazing, seven angels with seven plagues, which are the last, for with them the wrath of God is finished.

The following chapter in Revelation describes how each of the seven angels pours out a bowl of wrath on the earth, with each producing a plague on the earth.

In this brief survey of biblical passages describing apocalyptic signs in the heavens, we have seen themes repeated in passages from both the Old Testament and the New Testament. It is almost certain that many of these passages refer to the same events and hence ought to be correlated with one another. However, we must be cautious not to assume that similarity of languages implies an exact correspondence in every case. For instance, Ezekiel 32:7–8 appears to have been fulfilled, while other similar passages do not appear to have been fulfilled yet. One thing is certain: these signs, when fulfilled, will be astounding and perplexing. If these events were merely unusual but easily explainable, they hardly would constitute wondrous signs. The next section of this chapter gives an example that illustrates what I mean by this.

The Moon Turned to Blood

We have already seen that Joel 2:10 and Joel 3:15 speak of the sun and moon being darkened. Another passage in the book of Joel mentions that the sun shall be darkened, but it contains additional startling details. Joel 2:28–32 states,

> "And it shall come to pass afterward,
>   that I will pour out my Spirit on all flesh;
> your sons and your daughters shall prophesy,
>   and your old men shall dream dreams,
>   and your young men shall see visions.
> Even on the male and female servants
>   in those days I will pour out my Spirit.
>
> "And I will show wonders in the heavens and on the earth, blood and fire and columns of smoke. The sun shall be turned to darkness, and the moon to blood, before the great and awesome day of the LORD comes. And it shall come to pass that everyone who calls on the name of the LORD shall be saved. For in Mount Zion and in Jerusalem there shall be those who escape, as the LORD has said, and among the survivors shall be those whom the LORD calls."

Here wonders in heaven are mentioned, along with the sun being darkened. However, rather than the moon being darkened, it is described as turning to blood. While a very few might opine that this refers to the moon literally turning into blood, most people understand that this means that the moon will appear as blood. That is, the color of the moon will change to red. However, is the moon turning red the only understanding of this? Similar language is used elsewhere in an unrelated event. Revelation 16:3 states,

> The second angel poured out his bowl into the sea, and it became like the blood of a corpse, and every living thing died that was in the sea.

Notice here that the sea, like the moon in Joel 2:31, is turned to blood. However, Revelation 16:3 contains a further qualifier, namely that the blood will be like that of a corpse. The King James Version says that the sea "became as the blood of a dead man." We understand that when blood issues forth from the body it is red. However, what does blood that has been dead any period of time look like? As the blood cells die, blood rapidly darkens, becoming dark brown-black. Rather than the moon turning red as most people seem to think, could Joel 2:31 be describing the moon as being darkened? This makes sense in context of the many passages where both the sun and moon are darkened, such as Joel 2:10 and Joel 3:15. In this manner, Joel 2:31 describes an analogous event with both the sun and moon described as turning dark, but expressed in different ways.

Why did John in Revelation see fit to describe this concept more explicitly? Joel wrote to a Hebrew audience. These were people who had experienced the sacrificial system prescribed in the Levitical law. The altar and other furnishings in the Temple were not washed between sacrifices. Consequently, blood built up on them, darkly staining them. Those who originally heard and read Joel's prophecy probably would have made this connection and realized that what Joel meant was that the moon would be darkened. This is reinforced by the other Old Testament passages that describe the darkening of the moon along with the sun. However, John's audience was very different. If John wrote the book of Revelation in the last decade of the first century as many scholars believe, then the Temple had been destroyed for more than two decades. Furthermore, John addressed the book of Revelation to seven churches in Asia (modern day Turkey). At this late date, these congregations probably contained many Gentiles who had not experienced the sacrifices at the Temple. For that matter, living so far from Jerusalem, many Jewish believers in those congregations probably had not experienced the Temple sacrifices either. Therefore, many of John's immediate readers may not have made this connection of the description of blood as appearing dark rather than red as Joel's readers had. Without this qualifier, John's readers may not have caught this allusion to darkening.

There are two other passages that describe the moon as turning to blood. One passage is Acts 2:17–21, where the Apostle Peter quoted from Joel on the day of Pentecost. It reads,

"'And in the last days it shall be, God declares, that I will pour out my Spirit on all flesh,
   and your sons and your daughters shall prophesy,
and your young men shall see visions,
   and your old men shall dream dreams;
even on my male servants and female servants
   in those days I will pour out my Spirit, and they shall prophesy.
And I will show wonders in the heavens above
   and signs on the earth below,
      blood, and fire, and vapor of smoke;
the sun shall be turned to darkness
   and the moon to blood,
      before the day of the Lord comes, the great and magnificent day.
And it shall come to pass that everyone who
   calls upon the name of the Lord shall be saved.'"

The other passage is Revelation 6:12–14:

When he opened the sixth seal, I looked, and behold, there was a great earthquake, and the sun became black as sackcloth, the full moon became like blood, and the stars of the sky fell to the earth as the fig tree sheds its winter fruit when shaken by a gale. The sky vanished like a scroll that is being rolled up, and every mountain and island was removed from its place.

Notice here that the moon is said to be full. Interestingly, this description is missing from the King James Version, though it exists in every Greek New Testament manuscript of this part of Revelation. There does not appear to be any significance to the moon being full if it is turned red in color. However, if the intended meaning is that the moon is darkened, the moon being full is significant. The full moon is considerably brighter than any other lunar phase. Part of the reason for this is that the maximum amount of the moon is lit when the moon is full. But more significant is the physical principle of backscatter.

A higher percentage of light is reflected from the lunar surface when the light is reflected back in the direction of the incoming light, as at full moon. Thus, if the full moon is inexplicably darkened, the effect is greater than at any other phase.

We ought to interpret Scripture in light of Scripture. Therefore, we ought to consider as a whole passages that are correlated. In addition to Joel 2:10 and Joel 3:15, Luke 21:25 clearly links together the dimming of both the sun and the moon. As we have already seen, Isaiah 24:21–23 may as well. These passages almost undoubtedly refer to the same events. Thus it would appear that the dimming of the sun ought to be accompanied by the dimming of the moon, which strengthens the argument presented here that the statements about the moon turning to blood in Joel 2:31, Acts 2:20, and Revelation 6:12 refer to the moon being dimmed. Hence, the belief that the moon turning to blood refers to a red color that the moon will assume may not be correct.

The understanding that Joel 2:31 and Acts 2:20 refer to the moon turning red gained considerable attention in Christian circles in the years 2012 to 2015. Mark Biltz, a pastor from Tacoma, Washington, began talking about so-called "blood moons" a few years earlier. In 2013, John Hagee picked up on Biltz's ideas and published a very popular book, *Four Blood Moons: Something Is About to Change*. Biltz followed with his own book, *Blood Moons: Decoding the Imminent Heavenly Signs* the following year. However, Biltz's book did not sell nearly as many copies as Hagee's book did, nor did it garner the attention of Hagee's book. This was not surprising, because Hagee was already well known, but Biltz was not.

Biltz noted that the most common color of a total lunar eclipse is some shade of red. Biltz further noted that blood also is red, from which he concluded that the Bible verses that describe the moon turning to blood must refer to a total lunar eclipse or eclipses. Thus, Biltz began referring to total lunar eclipses as "blood moons." He did this without any exegetical or theological foundation. Prior to Biltz's usage, the term *blood moon* was an alternate name for the hunter's moon, the full moon after the harvest moon. The harvest moon is the full moon closest to the autumnal equinox, so the blood moon is a full moon appearing in early- to mid-autumn. However, by 2014 so much news reporting used Biltz's term to refer to a lunar eclipse that *blood moon* now sloppily has come to refer to any lunar eclipse.

Why do total lunar eclipses often appear some shade of red? A lunar eclipse occurs when the earth passes almost exactly between the sun and the moon, so that the earth's shadow falls on the moon. One might expect that the earth's shadow would be dark, but the earth's atmosphere bends light into the shadow. That is why the sky does not get dark immediately after sunset. As the sun slips below the horizon, the earth's atmosphere bends light around the earth's edge into the night. Only when the sun is well below the horizon does the shadow of the earth appear dark. At the same time that the earth's atmosphere bends light into the earth's shadow, molecules of air also scatter light. Shorter wavelength light (bluer light) is scattered more than longer wavelength light (redder light). The more scattered light is removed, leaving only shades of red in the earth's shadow (the *umbra*). This is also why sunset and sunrise often appear red. The appearance of the earth's umbra varies from eclipse to eclipse. While the most common color is some shade of red, some total lunar eclipses have appeared black. The earth's umbra during a total lunar eclipse almost never appears a shade of red that resembles blood. Normally, it is some shade of orange, closer to the color of copper or a pumpkin. Therefore, it is very imaginative at best to call a total lunar eclipse a "blood moon."

Total lunar eclipses are not that rare. Some years have no total lunar eclipses, but other years have two. Lunar eclipses can happen only during a full moon, but a lunar eclipse does not happen every full moon. This is because the moon's orbit around the earth is tilted a little more than 5° to the earth's orbit around the sun, so the earth's umbra normally passes above or below the moon when the moon is full. There are two times per year when the moon's orbit around the earth crosses the plain of the earth's orbit around the sun so that the earth's umbra can fall on the moon, producing an eclipse. We call these times eclipse seasons. An eclipse season is a little more than a month long. There is a full moon each month, so at least one lunar eclipse must happen each eclipse season, though there can be two lunar eclipses during a particular eclipse season. Eclipse seasons are nearly six months apart, and they occur progressively earlier each year.

Most lunar eclipses are not total. A total lunar eclipse happens when the moon is completely immersed in the earth's umbra. Often, the moon passes only partly through the earth's umbra, producing a partial eclipse. The earth's umbra in a partial eclipse appears black, while the uneclipsed portion of the moon appears very bright. Only if a partial eclipse is very close to being total can the human eye discern color in the umbra. The reason is that the uneclipsed portion of the moon is about 10,000 times brighter than the eclipsed portion. The much brighter uneclipsed portion of the moon causes the eye to constrict, shutting off light from all of the moon. Not enough light from the eclipsed part of the moon reaches the eye to trigger a response to our color-sensitive cone cells. During a total eclipse, the moon is so faint that our eyes dilate, and enough light enters our eyes to register color.

Relatedly, a penumbral eclipse happens when the moon fails to enter the earth's umbra but enters the earth's *penumbra.* The penumbra is the outer partial shadow of the earth. The eye compensates for the moon's slightly decreased brightness during a penumbral eclipse, so it is difficult for most people even to notice anything unusual during a penumbral eclipse. There is no noticeable color change during a penumbral eclipse.

Occasionally a total lunar eclipse happens during each of four successive eclipse seasons. This is called a tetrad, meaning four. Each total lunar eclipse of a tetrad will be nearly six months apart over an 18-month period. Biltz noticed that a tetrad of total lunar eclipses would happen during 2014 and 2015. What made this remarkable according to Biltz was that these eclipses coincided with Passover and Sukkot (i.e., the Feast of Tabernacles) during those two years. Biltz thought that it was the timing of this tetrad with the major feasts of Judaism that made this remarkable and hence constituted a sign as prophesied in Joel 2:31. Biltz found that over the past 500 years only three other tetrads coincided with these feasts, and each one had significance for Israel, at least in Biltz's estimation.[10] One such timed tetrad of lunar eclipses happened near the time the Jews were expelled from Spain in 1492. The next happened near the establishment of the modern state of Israel in 1948. The third tetrad happened near the occupation of Jerusalem by Israeli forces in 1967. On this basis, for a while Biltz was explicitly stating that the Rapture would occur on Sukkot either in 2014 or 2015. Biltz eventually backed off from such dogmatic statements, but he continued to make it clear in his pronouncements that this is what he had in mind. Hagee was more circumspect in his claims, stating merely that something significant was going to happen during the tetrad of 2014 and 2015. Hagee did not specifically say what this was, but many people understood that Hagee was implying that this significant event in either 2014 or 2015 would be the Rapture. Obviously, this did not happen, so what went wrong?

---

[10] Biltz and others conveniently ignore previous tetrads that similarly fell on the high Jewish feast days, because they did not coincide with any significant events.

That question has many answers, but we will examine just a few. There is nothing remarkable about the fact that lunar eclipses might happen on Passover and Sukkot. Both Passover and Sukkot must happen at full moon. But lunar eclipses must happen at full moon too. The eclipse seasons get progressively earlier each year in an 18.6 year cycle. Therefore, eclipse seasons coincide with the time of year that Passover and Sukkot occur about every 9.3 years (we must divide by two, because there are two eclipse seasons per year). Since there are approximately two months per year when eclipses can take place, we would expect that lunar eclipses will happen at the time of Passover and Sukkot about two-twelfths, or one-sixth of the time. This is exactly the frequency that lunar eclipses coincide with Passover and Sukkot. Remember, both lunar eclipses and the Jewish feasts must happen at full moon, so the eclipses that fall close to those feasts must fall exactly on those feasts. While most people are astounded with the supposed rarity of the coincidence of lunar eclipses and the Jewish feasts, those familiar with eclipses and the Hebrew lunisolar calendar are not amazed.

This entire discussion about "blood moons" sets the bar very low for prophecy fulfillment. Total lunar eclipses are not that rare, so Biltz appealed to the timing of these eclipses rather than the appearance of the eclipses as significant. But look at the statements that accompany Joel's prophecy. It speaks of wonders in heaven and earth, such as blood, fire, and columns of smoke. Where are these other apocalyptic signs? Biltz and Hagee have committed the error of snatching one phrase from Joel 2 and elevating it while ignoring the others. The eclipses themselves were not remarkable in appearance, so they hardly would have stood out as a sign to anyone. Appealing to the timing as being a sign is unremarkable when compared with the other signs mentioned in the same context by Joel. Most people who were excited about this nonsense did not even see any of the eclipses. Apparently they were satisfied with just knowing that somewhere an eclipse was going on.

Why did the Apostle Peter quote from Joel on the day of Pentecost? Supporters of the blood moon hypothesis would have us believe that a lunar eclipse six weeks earlier was on the minds of the people on that day. They believe that the Crucifixion was on April 3, AD 33 (see Chapter 8 for a discussion of the probable date of the Crucifixion). The moon rose at sunset while in eclipse that evening. Supposedly, the eclipse made the moon appear red, or as blood, as it rose that evening, which is why Peter quoted from Joel. However, that eclipse was partial, not total, and the eclipse was well past maximum when it rose over Jerusalem that evening. That is to say, the moon did not appear blood-like that evening, or at least it did not appear red because of the eclipse. One may invoke some other effect to cause the moon to have appeared red that evening, but then that would not require an eclipse. Nor is there any record that the moon did appear red that evening, so this entirely is conjecture.

To understand why Peter quoted from Joel, one must look at the context. The Holy Spirit had just been unleashed on Jesus' followers (Acts 2:1–4). The Holy Spirit moved the believers to speak, and their words were heard in the native tongues of the many people visiting Jerusalem (Acts 2:5–11). That was an incredible thing. In response to questions as to what was going on (Acts 2:12–13), Peter addressed the crowd (Acts 2:14–41). Peter began by quoting from Joel 2:28–32, though he omitted the last part of verse 32. The relevant part of Joel's prophecy is that God would pour out His Spirit in the latter days. Not all of Joel's prophecy was fulfilled that day. Indeed, it appears that parts of it are yet to be fulfilled. What people experienced that day was very exceptional, far more exceptional than the mere timing of lunar eclipses.

It is interesting that while Biltz and Hagee make much of Joel 2:28–32 and Acts 2:17–21, they rarely discuss the third passage that mentions the moon turning to blood, Revelation 6:12–14. This is probably because Biltz and Hagee are premillennial and pretribulational in their theology. By their eschatology, the Rapture must occur prior to the events of Revelation 6:12–14. Hence, Revelation 6:12–14 cannot apply to events leading up to the Rapture. However, by their similarity in language, it is likely that Joel 2:28–32, Acts 2:17–21, and Revelation 6:12–14 all refer to the same events. Hence, if the moon turns to blood after the Rapture, then the moon turning to blood cannot be a sign that precedes or accompanies the Rapture. In their attempt to sensationalize this, Biltz and Hagee have not clearly thought through their thesis.

I began this section with a discussion of why the moon being turned to blood may refer to the moon being darkened rather than the moon turning red. If this is the case, it undermines the blood moon thesis. One might respond that a totally eclipsed moon is darkened, so this fits either way. However, that would require major revisions in how the blood moon thesis is described. Furthermore, one might respond by noting that Revelation 6:12 describes the moon becoming like blood without the qualifier of it being like the blood of a corpse as Revelation 16:3 does. Since John was writing to an audience largely unfamiliar with the Levitical sacrifices, why did he not include this, if John meant to say that the moon would be darkened rather than turned a shade of red? There are many allusions in the book of Revelation to Old Testament prophetic literature, allusions of which many of the original readers would have been aware. For instance, there are parallels between the descriptions of God's throne in Revelation 4:2–11 and Ezekiel 1:4–28. In similar manner, Revelation 6:12 likely is an allusion not only to Joel 2:31, but other passages, such as Joel 2:10 and Joel 3:15. These other two verses in Joel are significant, because they, like Revelation 6:12–13, include mention of a drastic change in the stars. Also, there is a parallel structure in the description of the sun and moon in Revelation 6:12—the sun is compared to sackcloth, and the moon to blood. This parallel structure suggests that a similar thing is to happen to both the sun and moon. Hence, if the sun is darkened, so will the moon.

Concluding Remarks

Revelation 6:14 introduces another concept that we have not discussed yet. It says that, "The sky vanished like a scroll that is being rolled up." This appears to be a partial quote of Isaiah 34:4, which reads,

> All the host of heaven shall rot away,
>   and the skies roll up like a scroll.
> All their host shall fall,
>   as leaves fall from the vine,
>   like leaves falling from the fig tree.

The context here is the judgment of nations (cf. Isa. 34:1–2). In various parts of Isaiah and other prophetic books, judgment is pronounced on different nations. However, there is no list of nations presented in this portion of Isaiah. Rather, Edom alone is mentioned. It is not clear whether Edom is a stand-in for all nations as it were. Isaiah had spent considerable time condemning various nations, such as Babylon (chapter 13), Moab (chapters 15–16), Aram (chapter 17), Cush (chapter 18), Egypt (chapter 19), and Tyre and Sidon (chapter 23), but Edom alone is singled out here. A New Testament example of the judgment of the nations may offer a clue to what this may mean. Matthew 25:31–32 reads,

> When the Son of Man comes in his glory, and all the angels with him, then he will sit on his glorious throne. Before him will be gathered all the nations, and he will separate the people one from another as a shepherd separates the sheep from the goats.

One might think at first that this is judgment of nations as in Isaiah 13–19. However, from the rest of the passage (Matt. 25:34–46), it appears that this judgment will be directed toward individuals, not nations. That is, the nations mentioned in Matthew 25 do not refer to collective groups of people, but to individuals. To see this more clearly, try inserting the words "the people of" into Matthew 25:32 so that it reads, "Before him will be gathered (the people of) all the nations." Verse 31 ties this judgment to the time of Jesus' return to earth. The other passages that we have discussed here indicate cosmic signs will accompany His return, so since Isaiah includes cosmic upheaval in this passage, it could be that this is the same judgment that Jesus spoke of in Matthew 25.

Also notice in Isaiah 34:4 the reference to the stars falling from heaven. The only other verses that mention this are Matthew 24:29 and Revelation 6:13. Revelation 6:13 even compares the stars falling from heaven to a fig tree as does Isaiah 34:4, except that it is the fruit being blown loose, not the leaves that fall. This suggests that all three verses refer to the same event. However, these three verses may be difficult to collate, depending upon one's eschatology.

What does it mean for the stars to fall from heaven? It almost certainly means something very different today than it did in the past. Today we know that stars are far larger and more massive than the earth. Furthermore, the stars are at incredible distances from the earth. Therefore, stars do not fall to the earth. But that is viewing things within the modern understanding of stars. Appearance is the basis of whether an object is a star, biblically understood. Even today we recognize "falling stars," though they are not stars at all, but are meteors. A meteor results when a relatively small particle (a piece of an asteroid or a comet) moving at a high speed relative to the earth encounters the earth's atmosphere. At an elevation of about 100 kilometers, the atmosphere is thick enough so that its interaction with the debris burns the particle up. It is this burning up that we see as a brief, bright streak of light that resembles a star in appearance, albeit moving very rapidly. In a dark sky, meteors are not that rare, with several visible per hour. However, there are various times of the year when there are unusually high rates of meteors visible. We call these meteor showers. During a meteor shower, one can see dozens of meteors per hour. On rare occasions, there are very intense but very brief bursts of meteors during a meteor shower, so that thousands of meteors per hour are visible. We call this a meteor storm. Perhaps the prophesied stars falling from heaven will be an extremely intense, prolonged meteor storm unprecedented in history.

On the other hand, there is another possibility. Perhaps the stars falling from heaven refers to the absence of stars. Several passages prophesize that the stars will be dimmed. In this sense, the stars disappear and hence could be said to fall from heaven, though they do not literally fall.

The description of the heavens rolling up like a scroll is unique to Isaiah 34:4 and Revelation 6:14. Psalm 102:25–27 speaks of the heavens passing away, but not in apocalyptic terms. The subject is not judgment, but rather it is the constancy and eternality of God. God is eternal, but the creation is not. Psalm 102:25–27 is quoted in Hebrews 1:10–12.

Finally, in Luke 11:16 some people asked Jesus for a sign from heaven. This request came in the aftermath of Jesus casting out a demon from a man. Some said that Jesus did this by the power of Beelzebul, and others asked for a sign to test Him, but Jesus refused to grant such a sign. The Pharisees likewise had sought a sign from heaven concerning Jesus in Mark 8:11. (This does not appear to be the same event of Luke 11:16. This request was after Jesus fed the 4,000, albeit sometime later. It would seem that the miracle itself would have constituted a sign, but apparently they would not recognize this for the hardness of their hearts. Therefore, Jesus once again declined. Note that Matt. 16:1 is a parallel passage to Mark 8:11.) More detail is given here in that Jesus chided the Pharisees for readily acknowledging signs in the sky with regard to weather prognostication, but yet as leaders of the Jews they could not recognize the signs of the times. Instead, Jesus offered them the sign of Jonah, anticipating the three days that He would spend in the tomb. Ultimately, signs rarely seem to make a difference in the hearts of men. People demand evidence in the form of a sign, but when confronted with a sign, they often reject it anyway. Thus signs appear to be more condemning in nature than confirming.

Ultimately, we do not know now what form the prophesied signs in the heavens will take. Given the terrifying events that will accompany these signs, we can be certain that these signs will be recognized and they will be perplexing. Man undoubtedly will attempt natural explanations for them, but the people who believe those lame excuses will do so out of the desperation of rejecting the truth.

PART 3

Astronomical Questions and the Bible: How Scripture Confronts Recent Questions about Astronomy

The previous two parts of this book have explored what the Bible has to say about astronomy, at least in a direct sense. Part 3 now ventures into less certain territory. From time to time, questions arise, questions that the Christian may want to know the answers to, but the Bible itself appears to be silent about, or at best offers only hints.

Chapter 10 tackles a question that some see as coming from clues in certain Bible passages. Some passages of Scripture seem to suggest that the month was once 30 days long and that the year once had 360 days. One could use months that are any length that one wants, provided that one is not too particular about how well the month is in synch with the moon. That is the situation that we find ourselves in today with the Gregorian calendar. Ever since the Julian calendar reform two millennia ago, the Western calendar has not coincided with the moon's motion, and most of us don't care. On the other hand, a 360-day year would be problematic. Such a calendar would be useful in the short run of a few years, but after more than three or four years, the calendar would be hopelessly out of step with the seasons. Nearly everyone would find this unacceptable. So why do these Bible passages appear to use such a calendar? There are several possible answers to this question. One of the answers proposed is that the year originally was 360 days long, and that it consisted of 12 months, each month being exactly 30 days long, but that the calendar dramatically changed at some point. This is a drastic solution. Chapter 10 will investigate this proposal.

Likewise, the light-travel-time problem poses some significant questions. The universe appears to be mind-blowingly large, far larger than just a few thousand light years across. If the universe is only thousands of years old, how can we see objects that are millions, if not billions, of light years away? I frequently hear this question. It probably is the best argument for what we call old-universe creation, the belief that God created the universe billions of years ago. Since this is such a great problem for recent creation, it is not surprising that recent creationists have met the challenge and offered several solutions to this problem. In Chapter 11, I briefly survey the major proposed solutions to the light-travel-time problem. I conclude with what I think is the best solution—my own!

Part 3 concludes with a discussion of the question of whether life exists on other planets. Astrobiology, defined as the study of life elsewhere in the universe, now is a recognized science. There are many scientific meetings dedicated wholly or partly to astrobiology. There is a serious journal called *Astrobiology*, with papers on the subject in each issue. A person even can complete a Ph.D. in astrobiology. However, so far there is no evidence that life exists anywhere else in the universe. The only real biology associated with astrobiology focuses on extremophiles. Extremophiles are living organisms that exist in extreme environments on earth. They hope these are prototypes of possible alien organisms. But since astrobiology supposedly is the study of life elsewhere, astrobiology apparently is the one science for which there is no data. Instead, astrobiology mostly consists of speculation and conjecture about life elsewhere. And much of this speculation is based upon an evolutionary worldview. Well, two can play this game. While the Bible does not even remotely address the question of life elsewhere, we can apply biblical principles to address the question of extraterrestrial life.

CHAPTER 10

Does the Bible Teach the
Year Was Once 360 Days Long?

In Chapter 4, we discussed the calendar. The Jewish calendar is lunisolar, with generally alternating 29- and 30-day months and an intercalary month roughly every third year to keep in synch with the year's true 365¼-day length. The modern (Gregorian) calendar is solar and is designed to remain synchronized with the year's true length for millennia. However, there is a belief among many recent creationists that the year once had exactly 360 days and that the month was 30 days long. The reasons for this belief, the time at which the bases for calendars allegedly changed, and the manner in which they changed have many variations. In this chapter, we will examine some of these considerations and evaluate whether any are likely to be true.[1]

Reasons for Belief in a 360-Day Year

One motivation for belief that the month originally was 30 days and that the year was 360 days is what appears to some to be a cumbersome mismatch between the lengths of the day, month, and year. To some people, the fact that the year is not an integral multiple of the day or month and that the month is not divisible by the day appears to violate the description of the original creation as "very good" found in Genesis 1:31. However, are these people reading into that passage what their opinion of "very good" is? While the current arrangement may offend the mathematical sensibilities of some people, is it not a bit presumptuous to dogmatically assert that the current relationship

---

[1] Much of the content of this chapter has been published previously. See Danny R. Faulkner, "Was the Year Once 360 Days Long?" *Creation Research Society Quarterly* 49, no. 2 (2012): 100–08; and Danny R. Faulkner, "Analysis of Walt Brown's Model of a Pre-Flood 360-Day Year," *Creation Research Society Quarterly* 50, no. 4 (2014): 222–26.

between our timekeepers is somehow not "very good"? The pronouncement of the creation being "very good" stands in stark contrast to the ravages of sin that soon entered the world. That being the case, to be consistent, one ought to postulate that the mismatch in the natural units of time must have happened at the Fall, not at some later catastrophe. Yet proponents of the 30-day month and 360-day year theory generally do not make this case. Rather, most of them believe that the change occurred at the time of the Flood, not the Fall.

Another motivation is that many people believe that the prophetic year in the book of Daniel consists of 360 days. Note that the book of Daniel does not directly indicate that the prophetic year is 360 days. Indeed, the prophetic year is not even called a "year." The passage in question is the prophecy of 70 weeks in Daniel 9:24–30. The Hebrew of Daniel 9:24 literally reads "seventy sevens." The word for week is the word for seven, which is why this is translated into English as weeks. However, whenever a week of seven days is intended, the Hebrew word for seven is accompanied by יוֹם (yôm), the Hebrew word for day. Since the word yôm is lacking here, the intended meaning for "seven" is to refer to years. Daniel 9:25–26 states that 69 weeks of years shall pass from the order to rebuild Jerusalem until the Messiah is cut off. That is, there ought to be 483 (7×69) years between these two events. If we date Artaxerxes' order to rebuild Jerusalem (Neh. 2:1–8) to 444 BC, then we would expect the Crucifixion to be in the year AD 40. By the way, the Jews generally were aware of this understanding, because many of them, about the time of the birth of Jesus, knew that the time of the appearance of the Messiah was nigh. However, AD 40 is much too late for the Crucifixion. The most likely dates for the Crucifixion are either AD 30 or AD 33. But if the years intended here are 360 days long rather than 365¼ days long, then the duration is seven years less, 476 solar years, not 483 solar years. This precisely matches the AD 33 Crucifixion date.

Other than the fact that the length of time between Artaxerxes' order and the Crucifixion is wrong otherwise, what is the justification for suggesting that the prophetic year implied in Daniel 9:24–30 is 360 days long? Generally, an appeal is made to the book of Revelation. There is some justification in this, given the theological connection between the two books, as the quotations and allusions from Daniel found in Revelation are conspicuous. Several passages in Revelation mention a time that is (at least approximately) 3½ years in length. Many people think that this refers to one-half of the final seven weeks of Daniel's prophecy of weeks. Revelation 11:2 speaks of the city of Jerusalem being trampled by the nations for 42 months, 42 months being 3½ years. The next verse, Revelation 11:3, states that two witnesses will prophesy for 1,260 days. The implication is that these two time periods are concurrent, but this can be only if each of the 42 months is 30 days each. The figure of 1,260 days appears a second time in Revelation 12:6, and the period of 42 months occurs once again in Revelation 13:5, though it is not clear that these two lengths of time are coincident, or if they apply to the time period of Revelation 11:2–3.

One may properly ask why Daniel employed a year that was only 360 days long. There are several possible answers to this question. One possibility is the issue addressed by this chapter, the suggestion that the year originally was exactly 360 days longs with 12 exactly 30-day months. The catastrophe of the Flood supposedly changed the lengths of both the month and year. A variation of this theory is that through some future catastrophe during end times, the month once again will be 30 days and the year 360 days. While this might explain the use of 42 months and 1,260-day intervals in the book of Revelation, it does not explain the use of such a calendar during the time of Daniel. Daniel prophesied in the sixth century BC. Taking the very conservative (at least with regard to the antiquity of the date) Ussher chronology, the Flood was in the 24th century BC (other calculations place the Flood even earlier). Thus, if Daniel's use of a 360-day calendar was the result of the year actually being that long, one must question why Daniel used a calendar that was more than a millennium and a half out of date in his time. No one in Daniel's time observed such a calendar, and no one would have found such a calendar particularly useful.

There may be a much easier answer. In Chapter 4, we discussed how many ancient cultures approximated the length of the year in different ways. The point is, 360 is a very nice, round number, so it works well in estimating time. Even today some interest calculations are figured on a 360-day basis, but no one in the business sector thinks that the year actually is 360 days long. For instance, users of Microsoft Office should examine the Excel function DAYS360.

Measurement in 360 increments has certain advantages over base-10 measurements, such as the number of divisors. Ten is divisible by two and five, but 360 not only is divisible by two and five but also by three and, ultimately, other numbers, such as four, six, nine, and 12. Having so many divisors allows 360 to be easily portioned many different ways. As a strictly lunar calendar seems peculiar to us with our non-lunar calendar, so measurements in any other base than ten seems odd to us. There is nothing natural or obvious as to why we use base-10 mathematics. Most historians of math believe that we do so because we have ten digits on our hands (and toes on our feet!). Long division and multiplication are very cumbersome, but they are required with base-10 arithmetic. For the most part, this chore has been eliminated today with such widespread use of electronic calculators. But, until very recently, many computations were done with fractions, and this is where divisors are very helpful. Especially in science we are caught up with the supposed superiority of base-10 with the metric system, but for certain conversions, particularly small conversions, fractions work better.

Consider the common English standards of volume measurement. There are two cups in a pint, two pints in a quart, four quarts to a gallon, two gallons to a peck, and four pecks to a bushel. One can quickly see that there are 128 cups in a bushel (incidentally, this is base-2). Going the other direction, there is some base-3 (or alternately, base-6) involved. A ¼ cup contains four tablespoons, and there are three teaspoons in a tablespoon. Thus, one can quickly see that there are 48 teaspoons in a cup. Many younger people have difficulty with this, for they always have used calculators and hence never have been forced to use fractions in this way. But older people, particularly ones with much experience in cooking and baking, find that they can increase or decrease these measurements very easily to alter the size of a recipe. In similar fashion, people in pre-calculator times found it easier to work in bases other than ten. For instance, the UK monetary system was not decimal until 1971. Prior to that, it took 12 pence to make a shilling, and 20 shillings to make a pound sterling. There were three and six pence pieces (making up ¼ and ½ shilling, respectively). Many people today, particularly in the US, find this confusing in making change, but the British got by quite well for centuries with this system.

The ancient Babylonians had a base-60 number system (technically, it was a mingled base-6 and base-10). The Babylonians apparently introduced the division of the circle into 360°. This is important for several reasons. First, since their number system was already base-60, it required only multiplication by six to get to this figure. Second, the ancient Babylonians attached religious significance to the number six, though it is not entirely clear whether this came before or after the adoption of the use of the number six so much. (As an aside, the number of the beast, 666, appears to have a direct relation to the city of Babylon in the book of Revelation.) Third, the number 360 is very close to the number of days in a year, so at least over the short run, 360 days is a good approximation for the number of days in a year. Keep in mind that Daniel prophesied in Babylon, at the height of Babylonian power and influence. Within this culture, his readers would have understood this simplification without insisting that the year either was or had been exactly 360 days long.

As we saw above, the 69 weeks between the decree by Artaxerxes and the Crucifixion only works if one assumes that each of the time intervals encompassed by the 69 weeks consist of 360 days. While this is an approximation to the length of the year in the short run, after the 69 weeks, the difference between the exact length of the year and the approximation used amounted to seven years. No one claims that the year was exactly 360 days between these two events. Rather, Daniel used an approximation to the year as a motive in his prophecy. It would be more appropriate to say that each of Daniel's sevens was a time interval that closely approximated the year. However, the amount of actual days that occurred was precise (483×360 days = 173,880 days).

Again, the reasons for interpreting each seven of the 69 sevens of Daniel 9:24–30 as 360 days is that it brings concordance between the dates of Artaxerxes' decree and the Crucifixion and the suggestion that a similar scheme is employed in the book of Revelation. However, there is no good reason to interpret either or both of these prophetic passages onto the historical account of the Flood found in the book of Genesis.

Another biblical passage cited by proponents of the 360-day year is the Flood account. Genesis 7:11 records that the Flood began on the 17th day of the second month of the 600th year of Noah's life. Of course, nearly everyone is familiar with the account of rain for 40 days and nights (Gen. 7:12). Most recent creationists also are familiar with the statements in Genesis 7:24 and 8:3 that the water prevailed (literally, "was mighty") upon the earth for 150 days. That latter statement is followed by Genesis 8:4, which tells us that the ark rested on the mountains of Ararat on the 17th day of the seventh month. Assuming that this follows the chronology begun by Genesis 7:11 and that the statements of Genesis 8:3 and 8:4 refer to the same thing, many argue that 150 days here must exactly equal five months, implying a 30-day month. But is this the only possible meaning?

No, there are at least three other possibilities. First, there are a few assumptions listed in the previous paragraph. Those assumptions appear to be sound, but they are assumptions and thus ought to be clearly acknowledged.

For instance, the text does not require that verses 3 and 4 of Genesis 8 refer to the same events. They may, but we do not have, at present, a good enough understanding of the key relationships between the narrative's structure, syntax, and temporal progression to determine this with certainty. That is, we cannot necessarily conclude that the end of the prevailing of the waters coincided exactly with the same day that the ark rested upon the mountains of Ararat.

Second, there are several possible ways to understand this Genesis 8:4 date and the 150 days of Genesis 8:3. One is the aforementioned exact equivalence, with 150 days exactly equal to five months. However, there are other ways to understand this. We do not know what calendar was employed by Noah; all ancient calendars that we truly understand date from far later. At the time, Noah may have used an entirely different calendar, one with even a 30-day month, even though that month did not align with the synodic month and had some unknown mechanism to bring the calendar in line with the moon and the year. However, the calendar used by Noah probably is irrelevant. Since Moses authored the Pentateuch, it is far more relevant to know what calendar Moses used, because the original readers of the Flood account in the book of Genesis would have understood what they read in terms of the calendar then in use. The Bible does not directly instruct us as to the type of calendar that Moses employed. However, within the Pentateuch as elsewhere in the Old Testament, the term used for the first of the month is the term for a new moon (חֹדֶשׁ). A new moon coincides with the first of the month only on a lunar or solar-lunar calendar. This implies that Moses probably used a calendar of this type, thus eliminating the possibility that his calendar had 30-day months.

A third possibility is that the 150 days of Genesis 8:3 may be an approximation for the amount of time elapsed. Even today we approximate the length of the month by 30 days, as evidenced by so many financial and legal obligations stated in terms of 30, 45, 60, 90, and 180 days. Why are these numbers so often selected as opposed to, say, 10, 20, 50, or 100 days? Obviously, these are approximations to one, one and a half, two, three, and six months. In our overly litigious society today, the exact day count likely is to take precedence over an integral month count in tort matters, but the number of months undoubtedly is the intention. We do not know how many, if any, attorneys existed at the time of the Flood, but throughout history, an approximation of 30 days has been applied to the month, particularly when the number of months in question is low. In short, Genesis 8:4 contains a much more precise statement of time measurement than Genesis 8:3 does. That is, the length of time involved is five months (to the day), an interval of approximately 150 days.[2] This conclusion is consistent with the precision of the statements, does no harm to a straightforward reading of Scripture, and does not require that the original month was exactly 30 days long.

Supposed Support for the 360-Day Year from Ancient Sources

Many recent creationists who support the 360-day year claim that it is well established that many ancient cultures had a calendar based upon the year being 360 days long, but then abruptly changed to a calendar that was based upon a 365-day year. Supposedly, this was necessitated by a sudden change in the length of the year. However, this "fact" is not well established. Nearly all references for this trace back to Immanuel Velikovsky's 1950 book, *Worlds in Collision*. Velikovsky claimed that many ancient cultures once had a 360-day year but were forced to add an extra five days at some point because of some abrupt change in the length of the tropical year. For instance, Velikovsky stated,

> The Egyptian year was composed of 360 days before it became 365 by the addition of five days. The calendar of the Ebers Papyrus, a document of the New Kingdom, has a year of twelve months of thirty days each.
>
> In the ninth year of King Ptolemy Euergetes, or –238, a reform party among the Egyptian priests met at Canopus and drew up a decree; in 1866 it was discovered at Tanis in the Delta, inscribed on a tablet. The purpose of the decree was to harmonize the calendar with the seasons "according to the present arrangement of the world," as the text states. One day was ordered to be added every four years to the "three hundred and sixty days, and to the five days which were afterwards ordered to be added."
>
> The authors of the decree did not specify the particular date on which the five days were added to the 360 days, but they do say clearly that such a reform was instituted on some date after the period when the year was only 360 days long.[3]

Velikovsky apparently chose to interpret this addition of five days as the result of some actual change in the length of the calendar at that time, but that

---

[2] The eminent Hebrew scholar Umberto Cassuto takes this position. See *A Commentary on the Book of Genesis, Part II: From Noah to Abraham*, translated by I. Abrahams (Jerusalem: The Magnes Press, 1964), 43–44.

[3] Immanuel Velikovsky, *Worlds in Collision* (Garden City, NY: Doubleday, 1950), 331–32.

was not the case. In 1870, Samuel Sharpe translated the tablet that Velikovsky mentioned. Sharpe provided a translation of this passage in context:[4]

> So that the seasons also may do what is fit in every way according to the present arrangement of the world, and that it may not happen that some of the national festivals, which are held in the winter, should be sometimes held in the summer, in consequence of the star moving one day in four years, and that others of those now held in the summer, should be held in the winter in the future seasons, as had formerly happened to come to pass, from the arrangement of the natural year remaining of three hundred and sixty days and of the five days which were afterwards ordered to be added; from the first day the festival of the gods Euergetae being now carried forward, because of the four years, on to the five days added on before the new civil year; so that all men may know how the former defect in the arrangement of the seasons, and of the natural year, and of the decrees about the whole disposition of the pole, happened to be amended and made perfect by the gods Euergetae.[4]

Notice that the purpose of the decree is to implement the practice of a leap year, not to add five days to the 360-day year, for that was already being done. Velikosvky claimed that this marked the institution of adding five days upon the 360-day year, but he could do this only by quoting out of context and emphasizing what was not the purpose of the decree. Consider the editorial comments of Sharpe, who translated this stele:

> This Decree is valuable to us for other reasons besides its help to the study of hieroglyphics. It tells us of a proposal then made by the priests to reform the Egyptian calendar, at least, so far as it was used in fixing the days when the religious feasts were to be celebrated. Ever since the year B.C. 1322, in the reign of Menophra, probably the king better known as Thothmosis II., the Egyptian civil year had consisted of 365 days; and hence, for want of a leap-year, the new-year's day, and the feasts then celebrated, were always moving one day earlier every four years. This change, which must in every generation have been noticed, had now, by the help of the Alexandrian astronomers, been determined with greater exactness. The new-year's day, the 1st of Thoth, which ought to fall on the 18th of July, when the Dog Star is seen to rise heliacally, had now, in the ninth year of Ptolemy Euergetes, moved nearly nine months earlier and fell on the 22nd of October. This is well known from several observations recorded by the Alexandrian astronomers; and quite agrees—at least, as well as observations which depend upon eyesight and the weather can be expected to agree—with the information contained in this Decree, namely, that the Dog Star then rose heliacally on

---

[4] Samuel Sharpe, *The Decree of Canopus in Hieroglyphics and Greek with Translations and an Explanation of the Hieroglyphical Characters* (London, UK: John Russell Smith, 1870), 15–16.

the 1st of Payni. Calculating back from what we are told by Censorinus, our great authority on the Calendar, we should have supposed that was not the case till the next year, the 10th of Euergetes. The very small disagreement shows with what accuracy the heliacal rising of the star could be observed. However, the priests proposed to be no longer guided by this movable civil year in the arrangement of their feast days. How far their proposal was acted on we do not know. The change was not made by civil authority till the reign of Augustus, who first introduced the Julian mode of reckoning into Alexandria, in the year B.C. 25.[5]

Thus, Velikovsky takes a very different meaning from the text than that taken by the translator of the text. Using this approach, one could just as well claim that Julius Caesar's addition of leap year was required by some change in the actual length of the year during his lifetime or that the 1582 Gregorian calendar reform was necessitated by change that occurred in or shortly before 1582. Instead, both of these calendar reforms, along with the one that Velikovsky references, were required by earlier calendars that had failed to properly account for the true length of the tropical year.

Velikovsky continued,

In the fifth century Herodotus wrote: "The Egyptians, reckoning thirty days to each of the twelve months, add five days in every year over and above the number, and so the completed circle of seasons is made to agree with the calendar."[6]

Here Velikovsky quoted from the 1920 translation of Herodotus by A. D. Godley. The complete Herodotus passage earlier translated by George Rawlinson in 1858 read as follows, with the passage quoted by Velikovsky in italics:

The Egyptians, they said, were the first to discover the solar year, and to portion out its course into twelve parts. They obtained this knowledge from the stars. (To my mind they contrive their year much more cleverly than the Greeks, for these last every other year intercalate a whole month, but *the Egyptians, dividing the year into twelve months of thirty days each, add every year a space of five days besides, whereby the circuit of the seasons is made to return with uniformity.*)[7]

---

[5] Ibid., vii–viii.

[6] Velikovsky, *Worlds in Collision*, 332.

[7] Herodutus, *Herodotus—The Histories: The Second Book—Euterpe*, translated by George Rawlinson (New York, NY: Appleton and Company, 1858), http://www.neilixandria.com/index. php/Herodotus_-_The_Histories#The_Second_Book:_Euterpe (accessed Sept. 19, 2012).

Note that the quote in context places a different spin on the passage. What Herodotus was commenting on was the manner in which the Egyptians handled the fact that the year is not an integral multiple of the month, and Herodotus found the Egyptian solution to the problem preferable to that of the Greeks. Herodotus in no way stated that the Egyptians had to update their previously accurate calendar of twelve 30-day months because of some disjointed shift in the length of the year, as Velikovsky suggests.

In *The Natural History of Pliny* we read,

> I find also, that statues were erected in honour of Pythagoras and of Alcibiades, in the corners of the Comitium; in obedience to the command of the Pythian Apollo, who, in the Samnite War, had directed that statues of the bravest and the wisest of the Greeks should be erected in some conspicuous spot: and here they remained until Sylla, the Dictator, built the senate-house on the site. It is wonderful that the senate should then have preferred Pythagoras to Socrates, who, in consequence of his wisdom, had been preferred to all other men by the god himself; as, also, that they should have preferred Alcibiades for valour to so many other heroes; or, indeed, any one to Themistocles, who so greatly excelled in both qualities. The reason of the statues being raised on columns, was, that the persons represented might be elevated above other mortals; the same thing being signified by the use of arches, a new invention which had its origin among the Greeks. I am of opinion that there is no one to whom more statues were erected than to Demetrius Phalereus at Athens: for there were three hundred and sixty erected in his honour, there being reckoned at that period no more days in the year: these, however, were soon broken to pieces. The different tribes erected statues, in all the quarters of Rome, in honour of Marius Gratidianus, as already stated; but they were all thrown down by Sylla, when he entered Rome.[8]

Note that Pliny does not state that the year was 360 days long, but that it was "reckoned at that period no more days in the year."

As for the Persians, Velikovsky wrote,

> The ancient Persian year was composed of 360 days or twelve months of thirty days each. In the seventh century five *Gatha* days were added to the calendar.

> In the *Bundahis*, a sacred book of the Persians, the 180 successive appearances of the sun from the winter solstice to the summer solstice and from the summer solstice to the next winter solstice are described in these words: "There are a hundred and eighty apertures [*rogin*] in the east, and a hundred and eighty in

---

[8] John Bostock and Henry Thomas Riley, *The Natural History of Pliny* (London, UK: Henry G. Bohn, 1857), 159.

the west...and the sun, every day, comes in through an aperture, and goes out through an aperture....It comes back to Varak, in three hundred and sixty days and five Gatha days."

Gatha days are "five supplementary days added to the last of the twelve months of thirty days each, to complete the year; for these days no additional apertures are provided...." This arrangement seems to indicate that the idea of the apertures is older than the rectification of the calendar.[9]

For his first sentence, Velikovsky had this as a footnote and reference:

"Twelve months...of thirty days each...and the five Gatha-days at the end of the year." The Book of Denkart, in H. S. Hyberg, Texte zum mazdayasnischen Kalender (Uppsala, 1934), p. 9."[10]

Presumably, Velikovsky provided his own translation. Notice that the quote doesn't actually state when or why the practice of adding five extra days each year began. Velikovsky assumes that it was because of a catastrophic change in the lengths of the month and the year, but the text does not say this. The alleged catastrophic change is Velikovsky's hypothesis, but offering this as support is begging the question. Interestingly, elsewhere the *Book of Denkard* states,

Be it known that the solar year is of two kinds. Of these (two solar years) one is made up by the addition of days, the other by the addition of hours. The one that is made up by the addition of days consists of twelve months, each month of which is of thirty days. (When to these three hundred and sixty days) the five additional days, required for the course of the sun through the constellations during twelve months, are added the year becomes one of three hundred and sixty-five days. The five days which are over and above (the thirty days) of each month are placed at the end of the last month of the year. These five days are made up by the increase (in time of the solar year over the year of 360 days) and they are fixed after many calculations. According to such calculations these days are named (in the daily prayers recited on the last five days of the year).

Besides the sum-total of three hundred and sixty-five days there are six additional hours (to be taken into consideration). These hours have to be added every year. These additional (six) hours (for every year) make up one day for four years, ten days for forty years, one month for a hundred and twenty years, five months for six hundred years and one year for one thousand, four hundred and forty years. The time of six hours should be kept apart from (i.e. not to be added to) the last days of the year for many years, till (the hours) amount to (a definite period of time).

[9] Velikovsky, *Worlds in Collision*, 328–29.
[10] Ibid., 328, fn. 10.

This additional period (i.e. the intercalary month at the end of every hundred and twenty years) is fixed by calculations. And it (i.e. the intercalary month) is necessary for (the right performance of) Noruz, Mihragan, and other time-honored Jashans. Again the commencement of the year has been fixed by great kings from the first day of the year from the beginning of creation.[11]

Notice that the entire text in context clearly shows that the Persians knew that the year was 365 days long and that they added the extra five days to bring their twelve 30-day months into conformity with the actual year, as did the Greeks and Egyptians. More important, the last sentence indicates that this calendar had operated since the beginning of creation, thus contradicting Velikovsky's claim that Persian records indicate an initial 360-day year.

As for Velikovsky's use of the *Bundahis*, his first quote above is from chapter 5 of the *Bundahis*. The second quote is from a footnote of E. W. West, the translator of the *Bundahis* into English. Velikovsky clearly misinterpreted the meaning. Neither the text nor the footnote says that the five extra days were added to fix some drastic change from an earlier 360-day tropical year. In fact, elsewhere in the *Bundahis* the idea of the original year being 365 days is alluded to. In chapter 25, in dealing with the religious calendar of ancient Persia, we find this:

> On matters of religion it says in revelation thus: 'The creatures of the world were created by me complete in three hundred and sixty-five days,' that is, the six periods of the Gahambars which are completed in a year. . . .

> Again, the year dependent on the revolving moon is not equal to the computed year on this account, for the moon returns one time in twenty-nine, and one time in thirty days, and there are four hours (zaman) more than such a one of its years; as it says, that every one deceives where they speak about the moon (or month), except when they say that it comes twice in sixty days. Whoever keeps the year by the revolution of the moon mingles summer with winter and winter with summer.[12]

Note that the creation of the animals was said to have been accomplished in 365 days. This makes no sense if the originally created year actually was 360 days. Furthermore, this second quote explicitly states that the moon's orbit is 29½ days, and the last sentence firmly states that if one keeps a strictly lunar

---

[11] Peshotan Dastur Behramjee Sanjana, ed., *Denkard, The Acts of Religion* (Mumbai, India: Jeejeebhai Translation Fund), http://www.avesta.org/denkard/dk3s414.html (accessed June 2, 2012).

[12] Edward William West, *Sacred Books of the East*, vol. 5 (Oxford, UK: Oxford University Press, 1880), http://www.wisdomlib.org/zoroastrianism/book/the-bundahishn/d/doc4476.html (accessed Sept. 19, 2012).

calendar, the seasons will soon be out of cycle. When these statements are taken in total, it is very clear that Velikovsky's claim that the ancient Persians once had a 360-day calendar is not supported by the *Bundahis*.

This is the manner in which Velikovsky handled all of his support for his contention that the year was once 360 days long and then abruptly changed to 365 days. It is not clear whether Velikovsky merely misunderstood what he was reading or if he intentionally misrepresented the references to support his thesis. At any rate, careful analysis of the supporting references footnoted by Velikovsky reveals that the original sources in no way support the idea that the tropical year was once measured to be 360 days long. Unfortunately, many creationists have uncritically accepted and repeated Velikovsky's claims in this matter, and with no actual statements from antiquity concerning the tropical year actually being 360 days in length, the case for this is severely weakened.

Modeling Such a Change in the Month and Year

When did the lengths of the month and the year supposedly change? There is some disagreement on this, but the most common belief is that the change occurred at the time of the Flood. Others have suggested a post-Flood catastrophe such as the time of Babel or a supposed literal dividing of the earth at the time of Peleg. A change in the month from 30 days to the current 29½ days and a change in the year from 360 days to the current 365¼ days would require a change in at least two of the three natural units of time. How might this have happened? There is a shortage of specific proposals. The most straightforward approach would be to alter the orbital periods of the moon and earth while keeping the length of the day about the same.

Some people have suggested that impacts could have shortened the month and lengthened the year. Orbital dynamics can be counterintuitive. If a thrust, such as from an impact, is delivered to an orbiting body from behind (in the direction that the body is orbiting), the object's speed initially increases. The increased speed takes the orbiting body to a higher orbit. But higher orbits require less speed, so the body's orbital speed ends up less than before. Energy is not lost, because the decreasing speed is compensated by work done against gravity. Besides moving more slowly, a higher orbit has a greater circumference, so the body's orbital period increases. Conversely, a thrust, again from a collision, that opposes the orbital motion, results in a lower, faster orbit, so the orbital period decreases. Therefore, if the earth experienced a thrust from behind and the moon experienced a thrust opposing its motion, the length of the year would increase while the length of the month would decrease.

Exactly where on an orbit a thrust is delivered is important, which further complicates the situation. Orbits are rarely circular. Rather, orbits are ellipses with the body being orbited at one focus (this is called Kepler's first law). The point where a body orbiting the sun comes closest to the sun is perihelion; the most distant point from the sun on an orbit is aphelion. The corresponding points on an orbit around the earth are perigee and apogee.

If an accelerating thrust is delivered at perihelion/perigee, aphelion/apogee is raised, but perihelion/perigee remains unchanged. This makes an orbit more elliptical. Conversely, if an accelerating thrust is delivered at aphelion/apogee, perihelion/perigee is raised while aphelion/apogee is unaffected. This makes an orbit more circular. If instead of an accelerating thrust, a decelerating thrust is applied, the reverse of the above discussion is true. If a thrust is applied at any other point on an orbit other than perihelion/perigee or aphelion/apogee, then the computation gets more difficult.

So far we have considered thrusts applied exactly in the same direction or opposite the direction of motion. In a random situation, as with an impact, the thrust is not likely to be so perfectly aligned. Instead, the thrust will have a component in the direction of motion and a component perpendicular to the motion. The perpendicular component can change perihelion or aphelion distance, the orientation of the orbit, or the orbital plane, but it will not alter the orbital period, which is the desired effect.

Of crucial concern is the amount of non-orbital energy imparted to the earth during a collision. Since the moon is lifeless, we need not concern ourselves with this effect on the moon. A collision with the earth is a totally inelastic collision. In a totally inelastic collision, a portion of the energy of the impacting body could alter the earth's orbital period. However, the remaining portion of the impacting body's energy will be absorbed by the earth. The absorbed energy mostly will convert into heat, and this energy can be significant.

To illustrate this, consider a simple model requiring a near minimal amount of energy delivered to the earth to alter the year from an original 360 days to the current 365¼ days. Suppose that the earth's original orbit was not perfectly circular but less elliptical than today's orbit. An eccentricity about half of today's eccentricity but with the same perihelion distance would produce an original orbit with a 360-day period. An impact delivered at perihelion and directed forward in the earth's motion could produce today's orbit. Calculation shows that the required mass of the impacting body must have been about 0.28% the mass of the earth, a little more massive than Pluto. That is a large impacting body. If just 1% of the impacting body's kinetic energy were absorbed by the earth, and if that energy were uniformly distributed throughout the earth, the earth's temperature would have increased by about 100° C.

Of course, the energy absorbed would not be uniformly distributed, which means that the temperature increase in a large region near the vicinity of the impact site would have been far higher. The effects of such an impact would have been staggering, and this is for a minimum energy impact. For a less ideally oriented and timed impact, the energy released would have been far more—an actual impact likely would have been more energetic. Furthermore, this simple calculation assumed only a 1% conversion of the impacting body's energy going into heat on the earth. The actual percentage would have been far higher. Therefore, it does not seem likely that an impact could have so significantly increased the earth's orbital period without doing significant damage to the earth. If such a collision happened at the time of the Flood, the effects of the impact would have dwarfed the effects of the Flood. As the creation geologist Steve Austin once observed, it was *Noah's ark*, not *Noah's bunker*.

An alternate way to lengthen the year would be to shorten the day. That is, if the year remained the same but the day became shorter, the earth would rotate 365¼ times in the course of a year rather than 360 times. This could be accomplished either by applying a torque that sped the earth's rotation or by decreasing the earth's moment of inertia, perhaps by shrinking the earth in size. Application of a torque does not seem feasible, but a change in moment of inertia does.

In fact, Walt Brown, as part of his hydroplate model, has proposed that this happened. Today the earth is differentiated, with denser material, mostly iron and nickel, in the earth's core. However, if the earth originally was less differentiated, with more dense material near the surface, and if this denser material plunged into the earth's core at the time of the Flood, this would decrease the earth's moment of inertia. With no external torque applied, angular momentum must be conserved, and the earth's rotation rate would have increased, thus shortening the day. This portion of Brown's proposal is plausible. Many other Flood models, such as catastrophic plate tectonics, suggest a similar differentiation at the time of the Flood, so modern Flood models generally require at least a small decrease in the day's length at the time of the Flood.

This leaves the difficulty of changing the length of the month. This must happen by decreasing the moon's angular momentum, most likely by injecting energy at the right position, direction, and time to lower the moon's orbit. This is the reverse of the scenario of raising the earth's orbit discussed previously. Brown has proposed that impacts, primarily from debris ejected from the earth at the initiation of the Flood, provided the means to do this. However, there are problems with this scenario. Primarily, the amount of heat imparted to the moon would have melted much of the lunar surface. We see volcanic planes on the lunar surface (the lunar *maria*), but we believe that this was delivered from the moon's interior. Furthermore, the amount of melted material required by Brown's model greatly exceeds that of the lunar *maria*.

Conclusion

There is no biblical passage that states that the month once was 30 days and that the year once was 360 days. The idea that the month and year changed, most likely during the Flood, comes from the interpretation of certain biblical passages. However, those passages can be interpreted differently, so there is no scriptural requirement that there were such significant changes in the month and year. Many of the arguments for the year originally being 360 days allegedly come from nonbiblical ancient sources. However, this claim, in virtually all of its current iterations, can be traced back to Immanuel Velikovsky. Velikovsky either misunderstood or misrepresented the ancient sources, because none of the ancient sources actually claim that the month and year were altered. Finally, the proposed mechanisms that could have altered the lengths of the month and year are fraught with serious physical problems.

CHAPTER 11

Does the Bible Refute the Light-Travel-Time Problem?

Definition of the Problem

Perhaps the greatest difficulty for the recent creation model is the light-travel-time problem. The universe appears to be very large. For the general public, astronomers frequently describe distances by how long it would take light to travel those distances. The numbers used in these terms are a bit easier to visualize. For instance, the moon is about 250,000 miles (400,000 kilometers) from earth. Moving at the speed of 186,000 miles per second (300,000 kilometers per second), it takes light from the moon about 1⅓ seconds to reach the earth. Expressing the moon's distance in terms of light-travel-time probably is not preferable, because most people can grasp the moon's distance in miles or kilometers. If nothing else, they can picture the distance to the moon as being 30 times the earth's diameter or about ten times the earth's circumference.

However, the sun's distance is a bit more difficult to conceive. At a distance of 93 million miles (150 million kilometers), it takes light from the sun a little more than eight minutes to reach the earth. How do we go about visualizing that distance? Ninety-three million miles is too large for most people to grasp fully. That distance is 400 times the distance of the separation between the earth and moon. Or it is about 12,000 times the earth's diameter. Those comparisons are a bit difficult for most people to comprehend. However, a little more than eight minutes by light-travel-time is relatively easy, especially when compared to the moon's distance expressed as about 1⅓ seconds by light-travel-time. Rather than saying that Jupiter orbits the sun at a distance of about a half billion miles (nearly 800 million kilometers), we can say that Jupiter's orbit has a radius of about 45 minutes in light-travel-time. Saturn's distance from the sun is about

1½ hours. Pluto, one of the most distant objects in the solar system and farther from the sun than any planet, is only about 5½ hours away by light travel time.

But when we discuss stellar distances, human conception of such large numbers totally breaks down. Alpha Centauri, the closest star,[1] is about 275,000 times the distance between the earth and sun. That is nearly 26 trillion miles (more than 40 trillion kilometers). Numbers that large are virtually meaningless to the human mind. However, when expressed as a light travel time of a little more than four years, that distance is much more manageable. Therefore, we often say that Alpha Centauri is a little more than four light years away. A light year is about six trillion miles, or ten trillion kilometers. Of course, this is only the closest star (other than the sun)—other stars are much farther away. Most of the stars that we see at night are dozens or even hundreds of light years away. A few stars visible to the naked eye are a few thousand light years away.

However, we are just getting started with distances in astronomy. The Andromeda Galaxy is the closest galaxy that is similar in size to our own Milky Way Galaxy (there are some galaxies that are closer, but they are much smaller and fainter). The Andromeda Galaxy is a staggering 12 quintillion miles (about 20 quintillion kilometers) or more away. Obviously, no one can fathom that number, so we usually say that the Andromeda Galaxy is a little more than two million light years away. Even that number is difficult to grasp. However, as we noted already, we are just getting started here. The universe is filled with untold numbers of galaxies. Today, the most distant galaxies detected are on the order of 12 billion light years away. This upper limit is imposed by how faint distant galaxies appear owing to their tremendous distance, and by the inability of telescopes today to reach fainter light levels. With increased telescope size, we expect to push this frontier even farther. However, most astronomers, who think in terms of the big bang origin for the universe, suspect that we may be approaching the edge of the observable universe. On the other hand, the actual size of the universe may be far larger than the portion that we can actually see.

---

[1] Technically, Proxmia Centauri is slightly closer, but it is a very faint star and a member of Alpha Centauri system. This system consists of three stars. The two brighter stars orbit one another at an average separation a little more than the distance between Neptune and the Sun. These two stars appear as one to the naked eye, but even a small telescope reveals the two stars. Proxima Centauri is a very faint star that orbits the other two stars more than 500 times farther away than the two brighter ones orbit each other, and Proxima Centauri is much too faint to be visible without a large telescope.

The creation account and the genealogies in the Old Testament strongly imply that the world is only about 6,000 years old.[2] That would suggest that we ought not to see anything beyond a distance of 6,000 light years. Since the universe appears to be filled with objects that are much farther away than 6,000 light years, how can we see all of these objects? This is the light-travel-time problem. Creationists have proposed several solutions to the light-travel-time problem, and we shall discuss several of these in this chapter. However, before discussing those, we ought to ensure that we properly formulate the light-travel-time problem.

As it turns out, the problem is far worse than it initially appears! Most treatments of the light-travel-time problem discuss the situation as it now exists, 6,000 years after creation. That is, concentration is given to objects that are more than 6,000 light years away. These objects mostly are things outside of our own galaxy. However, these extragalactic objects generally are not visible to the naked eye. The average person will never see with their own eyes the objects that cause us such consternation with regard to the light-travel-time problem. Indeed, since the telescope was invented a mere four centuries ago, most people throughout history would not have recognized a problem. But that does not mean that the problem did not exist.

What did Adam and Eve see at the close of the sixth day, the day that God created them? As night fell, one would expect that they saw stars. Indeed, as we have seen, God made the stars two days earlier, on Day Four, and at that time He endowed the stars with several purposes. The stars could not have fulfilled their purposes if they could not be seen. Hence, anything more distant than two light days would have presented a problem. The solar system is only a matter of light hours across, so solar system objects (the sun, moon, and planets) easily would have been visible at the end of Day Six. However, since the closest stars are *light years* away, it would seem that they would not have been visible to Adam and Eve, at least not right away. There is no indication that the stars slowly blinked on over many years, and, as already pointed out, the stars could not have fulfilled their purposes right away if they had. Hence, by concentrating on the present time rather than the time immediately after the creation, we miss an important factor. Any solution to the light-travel-time

---

[2] This assumes that the Masoretic Text, not the Septuagint or the Samaritan Pentateuch, records the correct numbers in the Genesis 5 and 11 genealogies. The Masoretic Text is generally preferable as the closest reflection of the autographs, but that does not mean that it is superior in every instance. However, if one uses the Septuagint or the Samaritan Pentateuch, the date of creation is pushed back at most a millennium. This hardly is significant, nor can it solve the light-travel-time problem.

problem must answer the question not only of how we can see the most distant objects in the universe today, but also how Adam and Eve could have seen the closest objects outside of the solar system. It probably is more appropriate to concentrate on Adam's light-travel-time problem. If we can explain his light-travel-time problem, we probably can explain ours, too.

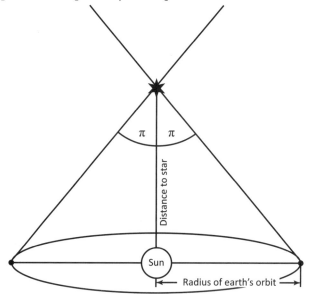

To illustrate the importance of this factor, let us consider one very poorly formulated solution to the light-travel-time problem. Some recent creationists have questioned the huge distances involved in astronomy. They correctly point out that the only direct method of measuring the distances of stars is by *trigonometric parallax*. As we orbit the sun each year, our vantage point changes from one side of the earth's orbit to the other. Hence, relatively close stars will seem to shift back and forth as we orbit the sun. We call this shift parallax. The amount of parallax depends upon the baseline (the radius of the earth's orbit) and how far away that a star is. The baseline is the same for all stars, so measuring the parallax allows us to use trigonometry to measure the star's distance. This principle has been used in surveying for centuries, if not thousands of years, to measure the heights and distances of objects on earth. You can illustrate parallax by holding your thumb up at arm's length and alternately viewing your thumb with one eye and then the other. The baseline is the distance between your eyes, and the parallax is the amount of shift that you observe. Until the 1990s, trigonometric parallax was limited to getting accurate distances out to about 60 light years. Since this is far less than 6,000 light years, this did not appear to be a problem in a creation that is only 6,000 years old. There are many other methods used in astronomy to measure distances much greater than what is possible by trigonometric parallax, but these methods are indirect because they require certain other assumptions. Some recent creationists then reasoned that since some of these assumptions may not be true, we really cannot trust these much greater distances.

There are many things wrong with this approach to solving the light-travel-time problem. For one thing, this approach usually does not clearly state that the vast majority of distances measured in astronomy are wrong, horribly wrong. But rather it is implied by kicking up dust and raising doubt. But raising doubt about assumptions is a far cry from actually refuting those assumptions. Many of those assumptions are well founded. Consider, for example, Cepheid variables. Cepheid variables in our galaxy appear to be very bright, pulsating stars. As they pulsate, Cepheid variables change size and temperature so that their brightness changes in a cyclical fashion. There is strong observational support for there being a relationship between the pulsation periods and the average intrinsic brightness of Cepheid variables. That is, the longer the period, the brighter the Cepheid variable is. There are also very good underlying physical reasons for seeing that this is so. We can see Cepheid variables in nearby galaxies. These Cepheid variables appear identical to relatively nearby Cepheid variables in our own galaxy, except that Cepheid variables in other galaxies appear much fainter. Their much fainter brightness is consistent with those Cepheid variables in other galaxies being millions of light years away. To dismiss cavalierly the assumptions of the Cepheid variable method is to undermine the underlying well-established physics and observations.

But more to the point, how does this approach solve Adam's light-travel-time problem? It does not, for it concentrates solely on the problem as it now exists. This supposed solution accepts the reality of stellar distances at least out to 20 light years while calling into question those much greater. But this would leave Adam and Eve with not a single visible star for more than four years and with very few stars for decades. We ought to focus our attention on solving Adam's light-travel-time problem first. If a solution cannot solve Adam's light-travel-time problem, then we ought to reject it. But any solution to Adam's light-travel-time problem probably can solve our modern light-travel-time problem.

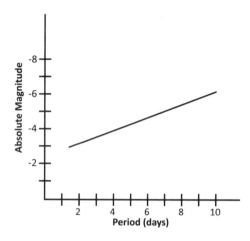

Mature Creation

For many years mature creation was the most popular solution to the light-travel-time problem. Its popularity undoubtedly is due to the fact that the co-founder of the modern creation movement, Henry Morris (1918–2006), espoused it. Morris reasoned that by its very nature, creation often is accompanied by maturity. One example is the plants made on Day Three. The plants were made mature, or at least they were mature by Days Five and Six. If the plants were not mature, then they could not have provided food for man and beast.

Another example is the feeding of the 5,000, the only miracle performed by Jesus recorded in all four gospels (Matt. 14:13–21; Mark 6:32–44; Luke 9:10–17; John 6:1–14). The only food available was five loaves of bread and two fish. Jesus blessed the food and instructed that it be passed out. The food miraculously multiplied so that all were well fed, with abundance afterward. We may surmise that there was nothing unusual about the multiplied bread and fish—they must have certainly looked and tasted like bread and fish, or else the people would have thought that the food was something else. However, it takes time to bake bread. The grain and any other ingredients must be grown, harvested, and prepared. The preparation presumably required for the bread to rise and certainly to bake, yet none of that appeared to have happened. Furthermore, fish take time to grow, but then they must be caught, transported to shore, prepared, and then cooked. All of this takes time, far more time than the brief moments that this miracle required. That is, Jesus created baked bread and cooked fish, food that is mature, yet did not go through all of the steps normally required for such things.

The same thing is true of the first miracle that Jesus performed, turning water into wine at the marriage in Cana (John 2:1–11). At some point during the wedding feast, there was no more wine. At His mother's bidding, Jesus instructed the servants to fill six large stone water pots and then to draw out the pots' contents and take it to the master of the feast. Once the master of the feast drank the wine, he called the groom over and commended him. The master of the feast commented that most people served the best wine first and then served poorer quality wine once people have had their fill, but this groom had saved the best for later. Wine normally comes from grapes, and grapes take months to develop. Once the grapes are harvested and pressed, the wine requires time to ferment. John 2:12–13 records that after the wedding feast, Jesus, His mother, His siblings, and His disciples departed to Capernaum, where he stayed a few days, whereupon Jesus went to Jerusalem, because it was Passover. Assuming that John 2:12–13 describe events that immediately follow those of John 2:1–11, then the marriage at Cana must have been in late winter or early spring. This is six months after the grape harvest, so any wine at that time normally would have been at least that old. Furthermore, wedding feasts typically went on for days, as much as a week. This miracle appears to have been at least a few days into the feast, and the master of the feast thought that wine was at least a few days old, because he thought that this exceptional wine had been held back until then (John 2:11). Yet the wine was only a few minutes old at most. If one had asked the master of the feast, or for that matter anyone else in attendance who had not observed the miracle as the servants had, they would have agreed that the wine had some age that was considerably older than it actually was. They would have assumed that the wine came about the way that wine normally comes about, but this omits the implied maturity of creation.

In similar manner, the creation account of Genesis 1–2 implies that Adam and Eve were created mature. For one thing, they were created with the ability of speech and hearing, as well as comprehension of what was said (Gen. 1:28–30; 2:15–17, 19–20, 22–24). Also, the fact that they were commanded to be fruitful and multiply indicates that they were of a certain level of maturity.

This would preclude Adam and Eve being created as infants and developing over time as people since have, since infants are incapable of engaging in and understanding speech; these skills require time to develop. If one could go back into time to the Garden of Eden a few days after the Creation Week and examine Adam and Eve, they probably would appear as particularly healthy young adults. We might estimate their ages as mid-20s, but we would be wrong. This is because we would be interpreting the recently-created Adam and Eve in terms of how we know that people normally develop.

Advocates of the mature creation solution to the light-travel-time problem argue that we must understand starlight in the same manner. It normally takes light many years to reach the earth from bodies outside the solar system. But since creation implies maturity of some type, this light must have been made with maturity built in. Just as Adam, Eve, plants, and probably animals in the Garden of Eden would have appeared as they normally do as the result of a somewhat long maturation process, distant stars must have appeared as they would have as the result of a somewhat long maturation process. This process of maturing amounts to light existing all along the path between the stars and us.

Since this would amount to God creating light between the stars and the earth, this often is called the "light created in transit" explanation of the light-travel-time problem. Supporters of this solution generally do not like this name, preferring instead to call it the "mature creation" solution. This distinction is not subtle. It is not just a matter of the light that we see. The light itself contains a wealth of information about involved physical processes, processes that we can infer from the contents of the light. This content consists of variations in brightness, spectral features, and polarization.

For instance, in 1987 there was a bright supernova in the Large Magellanic Cloud, a satellite galaxy of the Milky Way about 170,000 light years away. This was the first naked-eye supernova since the invention of the telescope more than four centuries ago. Astronomers observed the outburst of the supernova and its aftermath for years afterward. From the data collected, astronomers have pieced together what happened during the explosion and what transpired in the environment surrounding the explosion for years afterward. However, being much farther away than 6,000 light years, if God created this light in transit, then this event never happened.

Another example is eclipsing binary stars in the Andromeda Galaxy. A binary star actually is two stars orbiting one another via their mutual gravity. An eclipsing binary is a binary star oriented so that we happen to lie near its orbital plane. The stars orbit so closely that we cannot see the individual stars, so the system appears as a single star, even in large telescopes. However, as the two stars orbit, they periodically pass in front of one another. These eclipses cause the total light from the system to decrease and then recover. Astronomers measure the brightness of an eclipsing binary at all portions of the orbit to get a light curve. By studying the shapes and depths of the eclipses on the light curve, astronomers can determine many things about the stars. They can measure the sizes of the stars involved. Often astronomers can determine the masses of the two stars. They can determine the extent of tidal distortion of either star, resulting in stars that are not quite spherical, as most stars are. Sometimes there is evidence of dark or bright spots on the stars. There often is evidence of mass that is transferring from one star to the other.

There are thousands of eclipsing binary stars in our Milky Way Galaxy that are within only a few thousand light years. Of course, we have no difficulty in seeing these stars, even in a universe that is only thousands of years old. However, astronomers have found eclipsing binaries in the Andromeda Galaxy, a little more than two million light years away. The purpose of these studies was to use the measured sizes and temperatures of the stars involved to measure the distance to the Andromeda Galaxy. But if God created the light from the Andromeda Galaxy, then the light that we receive from it never actually left it. These eclipses never happened; they just appear that they did. But if the eclipses that we see never happened, then how can we say the objects that emitted the light even exist? Indeed, the vast majority of the universe is much too far away from us for light from much of it ever to have reached us by normal processes. Therefore, it is not necessary that much of the universe even exists, for things would be no different if it did not. God merely could have created the light and not the things that we think emitted the light.

But this means that much of the universe amounts to an illusion, and that would appear deceptive. Supporters of the mature creation solution reply that God has not deceived us, but that perhaps we have deceived ourselves, much like the master of the wedding feast of Cana. The master of the feast thought that the wine was older than a few minutes, but that was an inference that he made, not an implication that God gave. Or consider the fish that fed the 5,000. The fish presumably contained bones that normally took time to develop. After the meal, the disciples gathered 12 baskets of uneaten food (Matt. 14:20; Mark 6:43; Luke 9:17; John 6:12–13). Assuming that this food was distributed to the poor, anyone who received food relief from the gathered leftovers but had not heard of its source likely would have thought that the fish that he ate was perhaps a few years old. This has a certain logic, but why would someone think that wine and fish had some age much greater than mere minutes? It is because we have experience with the way things work. We know that wine, fish, and bread take time to grow and be prepared. We know that people, plants, and animals take time to grow. This is because we all have interacted with these sorts of things our entire lives. However, where is the similar relevant experience that tells us about distant astronomical bodies? There is none.

Furthermore, the fish, bread, and wine that people consumed was real. The animals that Adam named were real, and Eve certainly was real enough. However, the only thing real about light from distant astronomical bodies is the light itself. That is, if one wants to hypothesize that God created all this light from objects, that in no way requires that the objects actually exist. Since none of the light that most of these objects emit will never reach us—the heavens will one day pass away according to 2 Peter 3:10—then they serve no real purpose. This false history for the vast majority of the universe has prompted many creationists to look for other solutions to the light-travel-time problem.

C-Decay

In 1981, Barry Setterfield published a study of speed of light measurements made over the previous three centuries. He found a decreasing trend in the measured speed of light. The largest decrease in the speed of light was found in the earliest measurements. This suggested that the speed of light had been decreasing exponentially. When extrapolated into the past, this leads to the possibility that the speed of light may have been infinite or nearly infinite only a few thousand years ago. If the speed of light was so great during the Creation Week, then light from objects throughout the universe easily could have reached the earth, thus solving the light-travel-time problem. Because Setterfield's proposed mathematical description of the speed of light as a function of time is what we call an exponential decay, this solution to the light-travel-time problem often is called decaying speed of light. Physicists generally use the letter c to refer to the speed of light, so Setterfield's solution is frequently called c-decay (sometimes shortened to the clever phonetic letter sequence cdk).

Setterfield's proposal quickly gained attention, and many initial reactions were positive. However, other creation scientists soon began to evaluate it carefully. One objection raised right away was that the speed of light is not an arbitrary constant. Rather, the speed of light depends upon two fundamental constants of nature: the permittivity and permeability of free space. The permittivity of free space, ε_0, is involved in the attraction that charged particles have. The permeability of free space, μ_0, is involved in magnetic phenomena. The speed of light, in a vacuum, is

$$c = \frac{1}{\sqrt{\varepsilon_0 \mu_0}}$$

When the measured values of the permeability and permittivity of free space are inserted into this equation, it produces the correct value of the speed of light in vacuum. The presence of matter alters the permittivity and permeability, so within various substances we must use relative permittivity and permeability. Their values differ from the values in vacuum, so we do not use a subscripted zero in the relative values. Inserting the measured values of the relative permittivity and permeability of substances into the above equation correctly predicts the measured speed of light in those substances.

All of this is a result of Maxwell's theory of electricity and magnetism, one of the most successful theories of physics. (This theory is named for James Clerk Maxwell [1831–1879], the Scottish physicist who developed the theory). Given the robust nature of Maxwell's description of electricity and magnetism, it would seem that a change in the speed of light would require a change in ε_0 and μ_0 as well. However, ε_0 and μ_0 also are involved in the structure of matter. For instance, ε_0 regulates how strongly electrons are attracted to the nuclei of atoms. Even a very modest change in ε_0 would so alter chemistry and spectroscopy that its effects would be readily apparent. But the large change in c required by Setterfield's proposal would be anything but modest. In the estimate of many physicists, this would not work.

In response to this and other criticisms, Setterfield has refined his proposal over the years. Central to his ideas is zero-point energy. Zero-point energy is a hypothetical minimum energy that systems must have. Max Planck (1858–1947) first proposed the existence of zero-point energy in 1911 as part of his theory of quantum mechanics. Quantum mechanics, which was just being developed at the time, is the physics that best describes small systems, such as atoms and subatomic particles. Zero-point energy has been controversial ever since. Estimates of the possible value of zero-point energy vary widely. It has been invoked in fields as diverse the study of the smallest things that we can conceive to the largest, the universe. Setterfield has developed his theory so that the zero-point energy of the universe varies along with the other constants of nature. In this way, the effects of the changing physical constants are canceled. This effectively masks the changes except for the changing speed of light. It is not clear whether Setterfield's model can be tested experimentally. His critics have proposed observations that seem to contradict Setterfield's predictions. In response, Setterfield has altered his theory to account for the data.

What are we to make of the historical data that seem to indicate that the speed of light has changed? The earliest measurement of the speed of light is the most significant, because it has the greatest value, and thus drives the alleged exponential behavior more than any other datum. This earliest velocity of light measurement is attributed to Ole Rømer (1644–1710). Rømer did not actually compute a value for c. Rather, he was the first to demonstrate that the speed of light was not infinite. Despite Rømer's work, the finite speed of light was not fully accepted until more than a half century later.

Rømer's technique was to time when Jupiter eclipsed its satellites, but he particularly concentrated on the innermost satellite, Io, because its much shorter orbital period resulted in far more eclipses. Rømer found that Io's eclipses occurred earlier when Jupiter was closer to earth and later when Jupiter was farther from earth. He correctly attributed this to the time required for light to traverse the changing distance between Jupiter and the earth. If the speed of light were infinite, there would be no variation in eclipse timings. It was Christiaan Huygens (1629–1695), using Rømer's data, who first published a speed of light several years after Rømer's initial work. Interestingly, Huygens' value for c was about 28% *less* than that of the modern value. This is attributed to the fact that at the time the size of the earth's orbit was not well known. Note that if Setterfield had used the first *published* value for the speed of light, it would have contradicted his thesis.

Rather than use this first published measurement of c, Setterfield computed his own value, using Rømer's data, but the modern value for the size of the earth's orbit. Rømer had measured the diameter of the earth's orbit in terms of light-travel-time to be about 11 minutes, rather than the modern value of 16 minutes. Why the difference? Setterfield believed that Rømer accurately measured the light-travel-time diameter of the earth's orbit as it was during his lifetime. Thus, the discrepancy between the speed of light then and today is real and requires explanation.

But there is another explanation. Accurately timing when Io's eclipses occur requires an excellent clock. However, an accurate clock requires a regulator. We take this sort of thing for granted today, because we have so many reliable regulators to choose from. But in the 17th century, this was new and developing technology. Huygens is generally credited with making the first practical accurate clock. He used pendula and then oscillating springs to regulate his clocks.[3] This was about the time or shortly after that Rømer was doing his work, so it is doubtful that Rømer had access to such accurate time measurements. Again, the purpose of Rømer's work was not to measure the speed of light, but to demonstrate that the speed of light was not infinite. This did not require high precision. This does not mean that that Rømer was sloppy in his work, but rather that he simply used the rather imprecise time measurements available in his day. It is not proper to ascribe to Rømer technology that he did not yet have.

In 1849, Armand H. L. Fizeau (1819–1896) was the first man to measure the speed of light on earth. He reflected a beam of light off a mirror nearly nine kilometers away so that the total traveled distance was more than 17 kilometers. Dividing the accurately measured distance by the time elapsed for this trip yielded the speed of light. That time was only about 63 millionths of a second. How was Fizeau able to measure this time so precisely? Fizeau inserted a toothed wheel along the path of the outgoing beam so that the beam passed through a gap between two adjacent teeth. The reflected return beam traveled along the same path as the outward beam, so that it too passed through the gap. Fizeau arranged the apparatus so that he could view the returned beam. Fizeau used a handle to spin the wheel. At low speeds, Fizeau clearly could see the returned beam. However, as he spun the wheel faster, the outgoing beam passed through a gap, but the return beam was blocked by the adjacent tooth that had rotated into the place where the gap had been, so the observed beam was not visible. If Fizeau doubled the speed of the wheel, the beam again appeared, because the outgoing beam passed through one gap, while the reflected beam passed through the next adjacent gap. By measuring the rate of spin and knowing how many teeth the wheel had, Fizeau could measure the time interval.

---

[3] Likewise, for a modern mechanical clock, the regulator can be a pendulum or a spring. By contrast, an AC power clock relies upon the very regular 60 Hz oscillation provided by the utility. A battery powered clock often relies upon the vibrations of a very accurately shaped quartz crystal or a chip. Atomic clocks, the most accurate clocks that we have, are regulated by regular oscillations of electrons in particular atoms.

Fizeau's result was within 5% of the modern measurement of the speed of light. Being higher than today's adopted standard, if one takes Fizeau's measurement at face value, it suggests that the speed of light was higher in the past. Again, precision of time measurements may have been important. It would have been easy for Fizeau's measurement of the wheel's rotational speed to have been off by 5%. Other scientists worked to improve the accuracy. In 1862, Jean Bernard Leon Foucault (1819–1868) modified the Fizeau device to use a spinning mirror rather than a spinning wheel. He measured the speed of light to be slightly less (0.6%) than the modern accepted value. This was only thirteen years after Fiezeau's first measurement; did the speed of light change that much in that interval? Even supporters of cdk would say no. In 1869 Albert A. Michelson (1852–1931) used a similar device to measure the speed of light at only 0.07% greater than today's standard. Michaelson continued measuring the speed of light during the rest of his life with various collaborators. His measurements varied, but all were slightly higher than today's value. Near the end of Michelson's life, physicists began to use electro-optic devices to measure the speed of light, a practice that continues today. It is those experiments which provide the modern accepted speed of light.

Those who support cdk draw heavily upon Rømer's measurement of the light travel size of the earth's orbit, Fizeau's first measurement, and Michaelson's measurements that are greater than today's standard. They ignore or minimize measurements, such as Foucault's, that contradict their thesis. According to the cdk thesis, it would appear that Foucault's measurement was in error, but proponents refuse to allow similar conclusions about their favored measurements. Can opponents of cdk explain the general trend of decreasing measurements of the speed of light? Yes, they can. Trending is a subtle bias. If a person thinks that they know what the value of a measured quantity is, they tend to conduct their measurements and analysis in a manner that conforms to their expectation. This does not mean that the experimenter is dishonest or even consciously biased, but rather that he is human. If trending was at play, then it is not surprising that people measured values for the speed of light that gradually converged to the modern value.

At any rate, measurements of the speed of light seem to have been stable for much of the past century. How do supporters of cdk explain this? There are two possible answers. One answer is that an exponential decay eventually settles down to an equilibrium value, and so the speed of light has arrived or nearly arrived to that equilibrium value. Hence, no further change in the speed of light is expected. It is suspicious that the decay in the speed of light has reached saturation precisely at the time that we have the experimental precision to measure its change. In His sovereignty, is God that malicious? The second answer is to point out that the modern definition of units may have coincided with the constancy of the speed of light. In 1960, the unit of length (the meter) was redefined in terms of a wavelength of radiation from a particular electronic transition of the krypton-86 atom. A similar thing was done for time in 1967 when the second was redefined in terms of the period of oscillation of a particular vibration of the ground state of the cesium-133 atom. In a roundabout way, both of these definitions rely upon the speed of light. The relationship between the modern standard of length and the speed of light was more clearly established in 1983, when the meter was redefined in terms of how far light can travel in 1/299,792,458 second. This definition effectively defines the speed of light as 299,792,458 meters/second, rendering the speed of light a constant by definition. Therefore, it is not surprising that when one uses modern electro-optics devices to measure the speed of light, one gets this value (plus or minus experimental error).

This is a legitimate point—if the speed of light is changing. However, one cannot use mere suspicion that the speed of light is changing to argue that it is indeed changing. There are compelling physical reasons why the speed of light is a constant. Given the new modern standards of measurements, how could one show that the speed of light is not constant? The best approach probably would be to use a Fizeau/Foucault device for measuring the speed of light, being sure to use mechanical means rather than relying on standards as they are now defined. Apparently, no one has seriously conducted this sort of experiment in nearly a century.

Relativistic Solutions

In 1994, physicist Russ Humphreys published his famous white hole cosmology, quickly followed by a book entitled *Starlight and Time: Solving the Puzzle of Distant Starlight in a Young Universe*. Einstein's theory of general relativity postulates that time is not an absolute. Instead, the passage of time depends upon one's location in the universe. For instance, when one is farther from a very massive object, time passes more quickly than when one is closer to the massive object. This is because the gravity is stronger at the nearer location.

That is, gravity affects the passage of time. We call this time dilation. As strange as this sounds, time dilation has been confirmed by numerous experiments. One example is very accurate clocks at higher elevations on the earth, say at Denver, a mile high above sea level, record more time passing than a similar clock at sea level. Furthermore, time dilation is taken into account for a GPS system to work properly. If you have ever used GPS, you have employed relativistic calculations. The discrepancy between the clocks is miniscule, but measureable. Humphreys assumed that this well-established principle was at work early in the creation of the universe. Therefore, it is possible that if great differences in gravitational potential energies were present in the early universe, then great differences in time passage were possible.

More specifically, Humphreys proposed that God created the universe as a white hole with the earth near its center. A white hole is similar to a black hole, except that matter and energy fly outward rather than fall inward. The modern theory of black holes was developed in the 1960s by seeking a mathematical solution to the equations generated using general relativity theory. Theorists realized that there were two mathematical solutions, a black hole and a white hole. This often happens in physics—one solves an equation describing a situation and two different solutions emerge. Normally, the conditions of the problem being solved dictate which solution is the desired one. There is nothing physically wrong with a white hole, but there are at least two reasons why physicists do not expect white holes to exist in the universe today. First, a white hole is a contrived entity. That is, while we can envision natural processes that could produce a black hole, there is no natural process that could result in a white hole. Second, unlike black holes, white holes are unstable. Due to mass loss, white holes quickly evaporate and cease to exist. Therefore, if there were any white holes in the early universe, they long ago ceased to exist.

While theoretical physicists do not consider white holes any further, God creating the universe as a white hole can be quite appealing in a biblical cosmogony. In Humphrey's proposal, matter emerged from the white hole during the Creation Week and, at some point, enough matter escaped, causing the white hole to vanish. Since the earth was near the center of the white hole, it was among the last material to emerge. God made the stars of the universe on Day Four. As much of the universe emerged from the white hole, a great amount of time could have elapsed, allowing for the light of distant objects to have traveled across vast distances. However, due to time dilation, only a short amount of time—a day or two—elapsed on the earth. In this manner, light could have traveled great distance, while perhaps only a day transpired on the earth. Key to the Humphreys solution is the assumption that the perspective of the creation account is the earth. However, others believe that the proper perspective is the creation as a whole, not just the earth. In this view, the white hole solution amounts to an old universe.

Since 2008, Humphreys has taken his model in a new direction. He has developed a new metric, an equation describing space that fits Einstein's general relativity equations. This metric shows great promise for a cosmological model without a white hole. In this solution, a timeless zone develops that rapidly expands outward and then contracts back. Much of this happened on Day Four, when God made astronomical bodies. When the timeless zone contracted out of existence, the earth, which had been in the timeless zone, emerged, and since then time has progressed for the earth and much of the rest of the universe at about the same rate. Prior to this, much of the universe experienced time much more quickly than the earth, allowing for light from all the distant objects to reach the earth.

John Hartnett has pursued a solution to the light-travel-time problem using general relativity, but by using a somewhat different approach. Metrics used in general relativity theory normally have four dimensions, three for space and one for time. However, in the 1990s the Israeli physicist Moshe Carmeli began to develop a five-dimensional metric. This new metric includes the normal four dimensions, but adds a fifth dimension, called the space-velocity. The space velocity expresses the movement of galaxies as the universe expands. This eventually led to Carmeli publishing *Cosmological Special Relativity: The Large-Scale Structure of Space, Time, and Velocity* in 1997. Within a decade, Hartnett began to publish papers applying Carmeli's metric to the light-travel-time problem. This led to publication of his 2007 book *Starlight, Time, and the New Physics: How We Can See Starlight in Our Young Universe*. Hartnett has continued to develop his model to explain distant starlight, but he also has worked to explain other phenomena that we see in the universe as well.

Anisotropic Synchrony Convention

Jason Lisle first addressed the distant starlight problem in 2001, but he more fully explained his proposal in 2010. Lisle called his proposed solution the anisotropic synchrony convention, or ASC for short. We shall explain this name shortly. In modern relativity theory, one must adopt a synchrony convention that is related to the speed of light. Einstein chose to assume that the speed of light is the same in all directions. For instance, when measuring the speed of light as discussed earlier in this chapter, one measures the average speed that light must take during a round trip. The light travels to a mirror some distance away, and then the light travels back to the source by the same path, moving in the opposite direction. The average speed of light is twice the distance one way divided by the total elapsed time. Normally, one assumes the speed of light is the same either way, so the average speed of light is, by assumption, the speed of light. This assumption seems reasonable, but what if it is not true? Suppose that the speed of light is not the same either way, but that the speed *is infinite in one direction*. If the speed of light is infinite one way, then the speed the other way is *half* the average speed.

As strange as this assumption may seem, there is nothing necessarily wrong with it. Indeed, one must assume some convention in order to interpret the results of a speed of light experiment. In fact, it may not matter physically which convention you pick, for more than one convention could be valid. Most people would assume, as Einstein did, that the speed of light is isotropic, meaning that the speed of light is the same in every direction. Lisle calls this the Einstein synchrony convention. But Lisle proposed his anisotropic synchrony convention, where the speed of light is infinite in one direction. If the ASC is correct, then obviously light from all over the universe could have reached the earth on Day Four when God made the astronomical bodies. Within modern relativity theory, such a thing is possible. However, being possible and being correct are two different propositions. But again, it may not matter.

Lisle continues to develop the ASC. In the meantime, it has failed to attract much support in the creation community. To many people, this solution seems contrived, while many others simply do not understand it.

The *Dasha'* Solution

The word דָּשָׁא (*dāšā'*, commonly transcribed as *dasha'*) is a Hebrew word that that normally is translated as "bring forth" or "sprout." This word is used in Genesis 1:11 in reference to the vegetation that God made on Day Three. It was important that the vegetation be mature right away, for by Day Five and Day Six (just two and three days later), those plants were to provide food for both man and beast (Gen. 1:29–30). The plants could not have fulfilled this function unless they were mature. Many creationists would reply that God did make the plants mature. In fact, there is a common belief that God created everything *ex nihilo* during the Creation Week, implying instantly mature creation throughout. However, a careful reading of the creation account of Genesis 1–2 shows that this was not the case. Consider the detail of how God made man found in Genesis 2:7:

> Then the LORD God formed the man of dust from the ground and breathed into his nostrils the breath of life, and the man became a living creature.

Notice that God did not create man *ex nihilo*, but that God formed man from the dust of the ground. This dust probably existed since the dry land appeared on Day Three (Gen. 1:9–10). In turn, the matter from which the dry ground appeared probably was from *ex nihilo* creation of matter on Day One (Gen. 1:1). Thus, the matter from which God formed man probably was processed at least twice.

Or consider how God made plants on Day Three (Gen. 1:11–12). As previously mentioned, the Hebrew verb *dāšā'* appears in Genesis 1:11. A very different verb, יָצָא (*yāṣā'*), is used in Genesis 1:12. The two verbs are translated into English the same way, usually as "bring forth" or "sprout" (though in verse 12 *yāṣā'* properly is translated in the past tense). Neither of these Hebrew verbs suggest *ex nihilo* creation or even instantaneous appearance or maturity. Rather, they imply normal growth, albeit much more rapidly than we normally observe today. This suggests that the creation of plants on Day Four was a rapid growth and development to maturity, similar to what one today might see in time-lapse photography that is transformed into a short movie.

There are other examples. The appearance of dry land on Day Three implies a process, for Genesis 1:9 reads,

> And God said, "Let the waters under the heavens be gathered together into one place, and let the dry land appear." And it was so.

Notice the use of the word "gathered," suggesting a process whereby the waters moved to their appointed place, permitting the dry land to appear for the first time.

Another example is the creation of flying animals on Day Five and land animals on Day Six. From Genesis 1:20–21, one might think that God created birds *ex nihilo* on Day Five, and from Genesis 1:25 one might glean that God created land creatures *ex nihilo* on Day Six. However, Genesis 1:25 is preceded by Genesis 1:24, which reads,

And God said, "Let the earth bring forth living creatures according to their kinds—livestock and creeping things and beasts of the earth according to their kinds." And it was so.

The Hebrew verb translated "bring forth" is *yāṣā'*, the same verb that appears in Genesis 1:12 describing the creation of plants. Thus, the land animals proceeded out of the earth, similar to how the creation of plants was described in the Day Three account. Thus, God did not create land animals *ex nihilo*. Again, the language implies (rapid) process, not instantaneous creation. This concept is reinforced by Genesis 2:19, which reads,

> Now out of the ground the LORD God had formed every beast of the field and every bird of the heavens and brought them to the man to see what he would call them. And whatever the man called every living creature, that was its name.

The Hebrew verb translated "formed" here is again the word *yāṣā'*. This also is the same word that God used to describe His action in creating man from the dust of the earth in Genesis 2:7–8. Thus it appears that God used a similar method for creating land animals that He used in creating man. However, notice that Genesis 2:19 also includes birds (flying animals) with land animals in this description, so we see that God made birds from the ground as well.

It is clear that some aspects of instantaneous *ex nihilo* creation are within the creation account, such as the initial creation of matter in the beginning. But it also is clear that many aspects of the Creation Week involved processes. These processes are not to be confused with the supposed gradual, undirected, and naturalistic processes of evolution. Rather, the processes of the Creation Week were rapid and directed by God to complete His purposes within one week. Might we then expect that creation of astronomical bodies on Day Four involved processes? The Hebrew verb בָּרָא, almost universally translated as "create," does not appear in the Day Four account. Instead, the word עָשָׂה, translated as "make" in English, is used.[4] This leaves open the possibility that God made the heavenly bodies on Day Four from matter that He created *ex nihilo* on Day One. Of course, this does not preclude *ex nihilo* creation of astronomical bodies on Day Four; the language of the text could support either possibility.

---

[4] This verb is used elsewhere in the creation account, in Genesis 1:7, 11, 12, 16, 25, 27, 31; 2:2 (2×), 3.

But what of the light from astronomical bodies? As we have seen, the plants made on Day Three rapidly came to maturity. The reason for this was that the plants could not fulfill their purpose unless they rapidly matured. In a similar manner, the astronomical bodies could not fulfill their purposes unless they were readily visible. In this sense, the light had to be matured. No one suggests that the rapid development of dry land, plants, birds, land animals, and even man naturally happened. Rather, God rapidly and miraculously brought these things together to maturity. In a similar manner, why could not the light from distant astronomical bodies have undergone the same sort of process? Hence, the *dasha'* solution to the light-travel-time problem proposes that God rapidly and miraculously brought the light of distant astronomical sources to the earth on Day Four.

Some may see similarities between the *dasha'* solution and other proposed solutions to the light-travel-time problem. For instance, some may not see a great difference between mature creation and the *dasha'* solution. However, there are at least two distinctions. One is that the mature creation solution posits that God instantly created the light as to make the universe appear mature. However, the *dasha'* solution proposes that rather than creating the light mature, God rapidly matured the light emitted by heavenly bodies, much as He rapidly matured plants on Day Three. Another difference is that in the *dasha'* solution, light that we see actually left the objects in the heavens. In the mature creation solution, light that we perceive from much of the universe never did leave the objects that supposedly produced the light.

Some people may see similarity between the cdk and the *dasha'* solutions in that it appears that both rely upon a much more rapid speed of light in the past. However, the cdk solution assumes a physical mechanism whereby light traveled more quickly in the past. But the *dasha'* solution does not depend upon a physical mechanism. Furthermore, the *dasha'* solution does not require that the speed of light necessarily changed. It could be that some change in space accompanied this miracle. This could lead some people to see similarity between the *dasha'* solution and the relativistic solutions to the light-travel-time problem. However, the relativistic solutions attempt to answer the light-travel problem by appealing to a physical mechanism. What exactly happened is left unstated in the *dasha'* solution, because the events of Day Four were miraculous. Can we really understand the physical basis of miracles? Does that question even make sense?

Some creationists have criticized the *dasha'* solution on the basis that the *dasha'* solution appeals to a miracle. How is that a problem? Creation by its very nature is miraculous. Hence, the Creation Week certainly was filled with miraculous events. Of course, this supposes that we define a miracle as a departure from how things now work. However, the current sustaining of this present world by the power of God the Son's word (Heb. 1:3; Col. 1:17) did not yet exist until after creation. What we now call miraculous was the norm during the Creation Week. Searching for physical explanations for creation is no different than demanding physical mechanisms for other miracles, such as the Virgin Birth and the Resurrection of Jesus Christ.

Conclusion

The Bible appears to be clear enough that God created the world in only six normal days, and that the creation event was only thousands of years ago. The universe appears to be very large, much larger than just a few thousand light years. Hence, the light-travel-time problem is a serious problem for a recent creation. Creationists have responded with several solutions to this problem. More solutions almost certainly will appear. Which one is correct? Are any of them correct? We may not know this side of glory. One should not be discouraged by the multiplicity of proposed solutions. Instead, we ought to be encouraged. Perhaps in our frustration we overlook a very important point: does it matter? A God who could call all of creation into existence must be very powerful (Rom. 1:20). To a God so mighty, is the light-travel-time problem really a problem? Compared to the great miracle of the creation of the world, the light-travel-time problem seems miniscule.

CHAPTER 12

Does the Bible Address the Idea of Extraterrestrial Life?

The existence of life elsewhere in the universe is a belief widely accepted in our society. The sun is just one of a couple of hundred billion stars in our Milky Way galaxy. We now believe that there must be at least 100 billion other galaxies. In recent years, we have discovered many planets orbiting other stars, so it appears that extrasolar planets probably are common. Therefore, the number of planets in the universe must be mind-bogglingly large. However, everyone agrees that the vast majority of planets could not support life. Many stars that may have orbiting planets are unsuitable for various reasons. For a planet to support life, it must be the proper distance from its star—if the planet is too close to the star, it will be too hot, but if it is far away from the star, it will be too cool. The planet must have sufficient mass so that its gravity can maintain an atmosphere, but not too massive so that it has the wrong kind of atmosphere. These are just a few of the factors that must come together for a planet to be a suitable habitat for life. Some people reason that if only a tiny fraction of those planets are hospitable to life, then the number of planets where life exists must be huge. Therefore, even intelligent life may be common elsewhere in the universe.

What is the Christian to think about the possibility of life on other planets? Like any other subject, we ought to turn to Scripture for answers. Lest you think that you somehow missed that passage, the fact is that the Bible does not directly address the question of whether any life, let alone intelligent life, exists elsewhere in the universe. However, like so many other questions in life that the Bible does not directly address, we may use biblical principles to guide us in finding answers.

We know that spiritual beings—angels and demons—exist, and that they dwell in the heavenly realm. However, these creatures are not the sorts of things that we are discussing here. Rather, the question at hand is whether soulish beings, similar to humans, might exist somewhere else in the universe. This is what most people think is the important question, and that is usually what people have in mind when they ask about space aliens.

The Assumption of Evolution

We began this chapter with a discussion on how many planets that are suitable for life might exist in the universe. So far, we have not conclusively found any other planets where life could exist. However, we have found many planets where life clearly could not exist.[1] Therefore, the evidence to date shows that truly earth-like planets must be very rare, if they exist at all. Any claims that earth-like planets are common are based more upon wishful thinking or expectations from a particular worldview than actual data. Still, many scientists insist that even if the probability is low for any particular planet to be hospitable to life, the incredibly huge number of planets in the universe makes it likely that life, even intelligent life, must be common elsewhere.

Deeply hidden in this line of reasoning is the assumption that life arises spontaneously wherever the conditions are right. That is, life has a naturalistic origin. Of course, this presupposes that life arose on earth without the aid of a Creator. The biblically-minded Christian ought to recognize that this contradicts the clear teaching of Genesis 1 and 2. It is inconsistent to think that God created man and living things on earth but that life elsewhere in the universe arose naturally. Therefore, if life exists elsewhere, then it is because God created that life. God had His purposes in creating us, and so if God created life elsewhere, He would have had His purposes for creating that life too. Therefore, to the Christian, the question of life on other planets is a theological one.

As it turns out, science also has weighed in on the question of whether life exists elsewhere. There are least three lines of scientific evidence that suggest that life does not exist elsewhere. First, as just mentioned, scientists have not yet found any earth-like planets. As of 2016, astronomers have found approximately 2,000 extrasolar planets, but that number is sure to increase.

---

[1] Admittedly, there is a selection bias in that extrasolar planet detection techniques tend more often to find larger planets rather than smaller planets and planets orbiting closer to rather than farther from their stars.

The vast majority of these planets obviously cannot sustain life. Many are far too close or too far from the stars they orbit for liquid water to exist on their surfaces. Liquid water appears to be absolutely essential for life. Many extrasolar planets found so far are very massive and thus probably contain the wrong gases in their atmosphere for life to exist on them. Only a very few of the extrasolar planets found so far are within the critical habitable zone with the proper distance from their stars. Of those, even fewer are of the right size. The very few candidates remaining must be interpreted in terms of the most liberal assumptions in order for them to support life. No one can definitively say that any extrasolar planets found thus far are suitable for life. A critic might complain that we do not yet have a complete data set. That is true, but when is the data set ever going to be complete? We cannot conduct science on the basis of what we might discover tomorrow, or next week, or next year. We must work with the data that we have now. And the data now available shows that there is no planet other than earth where life is possible. With new data, that might change, but it might not; we simply do not know.

Second, science has repeatedly shown that the law of biogenesis is true. That is, life comes only from living things. Abiogenesis, life arising spontaneously from nonliving things, has never been observed. Belief in the naturalistic origin of life requires assuming something that has been disproven scientifically. There are only three possible explanations for the existence of life on earth:

1. Life has always existed, which requires that the universe has always existed. There are many reasons to believe that the universe had a beginning at a finite time in the past. At any rate, very few scientists today believe in an eternal universe.
2. Life came about through a supernatural process of creation.
3. Despite clear evidence to the contrary, life on earth came about through abiogenesis. That is, life arose on earth through natural processes from nonliving things sometime in the past, despite abundant evidence that abiogenesis does not occur. That would make abiogenesis a rare anomaly, suggesting that life may be unique to the earth.

Some may insist that the second option is unscientific. And indeed it is, because the second option invokes a non-natural process, and the scientific method is applicable to natural processes only. But this overlooks the fact that the other two options are also unscientific—option three assumes a process that is contradicted by abundant scientific evidence, and option one opposes the overwhelming scientific consensus that the universe has a finite age. Only extreme bias against any possibility of the supernatural would lead one to reject the second possibility. The fact that none of the three options are scientific underscores the fact that the origin of life is not a scientific question. Therefore, as far as science has been able to demonstrate to this point, life does not exist elsewhere.

Third, science has the potential to prove the existence of life elsewhere in the universe by direct means, but thus far it has failed to do so. The direct observation of life could be done one of two ways. One method would be to produce evidence of alien visitation on earth, either by directly observing a visitation or by artifacts left behind on previous visits. While many people believe that aliens regularly visit the earth today or that aliens helped ancient man construct certain structures (such as the great pyramids of Egypt) in the past, there is no good evidence for either of these beliefs. In 1950, the famous physicist Enrico Fermi took notice of this fact. He reasoned that if life is common in the universe, then it is probable that civilizations more advanced than our own already developed long ago in our galaxy. Once a civilization develops space travel, colonization of other planetary systems ought to follow relatively quickly, but certainly within a few million years. Therefore, at least one of these civilizations should have already reached the earth, so where are the aliens? This lack of evidence for aliens is often called the Fermi paradox, or the Fermi-Hart paradox, also crediting Michael H. Hart who wrote about the lack of evidence for aliens 25 years after Fermi. Of course, this is a paradox only if one believes that life is common in the universe.

The second method of detecting alien civilizations is via SETI, the Search for Extra Terrestrial Intelligence. Man has broadcast radio signals for about a century. Therefore, alien civilizations within a hundred light years of the earth could eavesdrop on those radio transmissions if they used a sensitive detector attached to a large radio telescope pointed toward the sun, around which the earth revolves. Likewise, we can reverse this process and listen in on alien transmissions by pointing our radio telescopes toward other stars. This has been the technique of SETI, the first attempt of which was done in 1960. Over the years, advances in computer and electronic technology have greatly increased the efficiency with which SETI can be done. Today SETI has become a nearly continuous process. What has the abundant accumulation of SETI data shown? There has been silence. Once again, the scientific data strongly imply that extraterrestrial civilizations do not exist.

Despite the evidence briefly discussed here, most scientists seem to persist in the belief that life must be common in the universe. This belief opposes the best evidence, so what is the reason for belief in extraterrestrial life? Alas, it is their commitment to belief in philosophical naturalism, which requires that biological evolution must be true. They understand that if life is unique to the earth, then that strongly implies that the earth must be special. The unique and special status of the earth easily leads to the consideration of creation, which many scientists are not prepared to do.

Theological Considerations

Of course, to the Bible-believing Christian, the possibility of creation is more than just a possibility—it is reality. Some Christians ask that if God created life on earth why He couldn't have created life on other planets. God certainly *could* have created life on other planets. However, there are certain theological issues that arise. While the Bible may not be geocentric, placing the earth at the center of the universe, earth and man do appear to be the center of God's attention. Life on other planets would undermine this.

If God created beings like us on other planets, then are they sinful, fallen creatures in need of redemption as we are? If so, then what was the cause of their fallen state? Was it because of Adam's sin? If it was for Adam's sin, then Christ's redemption on earth might suffice to atone for their sins as well as humanity's sins. However, to inhabitants of other planets, we and our ancestors Adam and Eve are the aliens. The gospel message on another world might begin, "A long time ago, in a galaxy, far, far away...." Scripture makes it very clear that we inherit our sinful nature through our ultimate ancestor, Adam (Rom. 5:12, 18–19; 1 Cor. 15:21–22). But aliens are not the descendants of Adam, so how was Adam's transgression imputed to them? No, it does not appear that Adam's sin can account for the sinful state of alien races.

If alien races did not obtain their sinful natures from Adam's sin, then what was the source of sin in their worlds? It must have been because the original beings in their worlds, sinned similar to how Adam and Eve sinned in the Garden of Eden. But that would require that Jesus be born of a virgin on another planet, live a sinless life, lay down His life for atonement of their sins, rise from the dead, and then ascend from that world to go on to other worlds to repeat the process. Besides the apparent cheapening of Christ's atoning work, the Bible makes it clear that Jesus did not go to another planet when He left the earth, but that He ascended to Heaven to the right hand of the Father (Matt. 26:64; Luke 22:69; Acts 2:32–33; 7:55–56; Eph. 1:20; Col. 3:1; Heb. 3:1; 8:1; 10:11–12; 12:2). Therefore, Jesus Christ, quite literally, died *once* for atonement of sins (Heb. 10:11–12).

Some might reply that beings similar to us living on other planets did not fall into sin. However, Romans 8:18–25 makes it clear that the taint of man's sin has affected the entire creation. That is why not only the earth, but the heavens too, will pass away (2 Pet. 3:10–11) and why there will be a new heaven and a new earth (Isa. 65:17; 66:22; 2 Pet. 3:12; Rev. 21:1). Not only do we require new bodies to fully overcome the taint of sin (1 Cor. 15:35–49; Phil. 3:20–21), but the entire universe must be redeemed from the blot of man's sin as well. Therefore, other planets have been affected by the consequences of sin.

We sometimes are loose in equating this defilement with the curse. Genesis 3 placed a curse upon two things: the serpent (verse 14) and the ground (verse 17). Man himself is not cursed, but man does live with the consequences, or taint, of his sin. This affect of man's sin stems from the headship of Adam (1 Cor. 15:21–22) in that death entered the world through Adam's disobedience. Many Christians erroneously conclude that animals were cursed in the Garden of Eden as a consequence of Adam's sin. The text for this supposedly is Genesis 3:14, where God placed a curse upon the serpent. That verse reads,

The LORD God said the serpent,

"Because you have done this,
   cursed are you above all livestock
   and above all beasts of the field;
on your belly you shall go,
   and dust you shall eat
   all the days of your life."

The reasoning is that since the serpent is cursed *above* (i.e., more than) all other animals, all other animals are cursed too, but just to a lesser degree. This is an unfortunate construction in English, for this is the implication that many people take from it. However, an identical construction in Hebrew occurs in verse 1, which reads,

Now the serpent was more crafty than any other beast of the field that the LORD God had made.

Just as this verse need not indicate that all creatures are inherently crafty (which is certainly not the case), so too Genesis 3:14 does not necessarily indicate that all animals are cursed. Also, we may note that there is a theological problem in charging that all animals are cursed. Would God, who, under the Mosaic Law, required perfect, unblemished sacrifices (Exod. 12:5; 29:1; Lev. 1:3, 10; 22:18–25; Deut. 15:21; 17:1; Ezek. 43:23; Mal. 1:8), find acceptable the offering of a creature that had been placed under a curse? Ironically, Malachi 1:14 condemns under a curse (note the Hebrew root אָרַר—the same root used in Gen. 3:14) anyone who would presume to offer a defective animal to the Lord![2]

Accordingly, we may conclude that animals in general are not cursed. However, they do suffer the taint of man's sin, for they are under man's headship.

Even if beings like us do not exist on other planets, might animals, plants, and "simple" life exist elsewhere? First, keep in mind that plants are not alive in the biblical sense. The Bible never speaks of plants living or dying. Rather, they bloom, but eventually they whither and fade. Something cannot die if it is not alive in the first place. Leviticus 17:11 tells us that life is in the blood. This amounts to a simple biblical definition of life—it would seem that one must possess blood to be alive. Since plants do not possess life, they are not alive. This does not mean that one cannot speak of plants being alive in some sense, but that they do not meet the biblical standard for life. Life is a precious thing, which is why all creatures were vegetarian before the Fall (Gen. 1:30). Eating animals would have required death prior to death entering the world through sin. However, eating plants was good, because plants are not alive in a biblical sense. In a similar manner, microorganisms and many other creatures, such as corals, do not possess blood, so they are not alive in a biblical sense either.

What would be the purpose of animals or other creatures on other planets? Man was given dominion over the creation (Gen. 1:28; Ps. 8:5–8). This dominion partly means that we can use the creation for our needs. Therefore,

---

[2] We see another noteworthy example in Genesis 3:16 which, though different syntactically from Genesis 3:1 or 3:14, nevertheless conveys a similar point conceptually. It reads, "To the woman he said, 'I will surely multiply your pain in childbearing; in pain you shall bring forth children. Your desire shall be for your husband, and he shall rule over you.'" Applying the same reasoning here as that which is commonly applied to Genesis 3:14 by those who maintain that all animals are cursed, it would appear that the original creation, without sin, would have involved pain in childbearing, albeit lessened from what it is now. However, most of the people who would insist that verse 14 involves a curse on all animals would insist that verse 16 refers to the introduction of pain in childbirth, not an increase in that pain. This is inconsistent. It is more likely that verse 16 refers to the introduction of pain in childbearing and verse 14 does not mean that all animals are cursed, but that the serpent alone was cursed.

other living creatures on earth are to serve man. The dominion mandate gives man the authority to rule the earth. If there were living creatures on other planets but no human-like creatures to have dominion, there would not be anyone in charge of these creatures. What then would be the purpose of living things on other planets?

The Objection of Wastefulness

Some people argue that if God did not create intelligent life elsewhere, then He wasted a large amount of space. The universe appears to be billions of light years across. Within the universe there are untold billions upon billions of stars. It now looks as if a significant fraction of stars have orbiting planets, so the number of planets likely exceeds the number of stars. Surely, some people reason, God must have made intelligent life elsewhere to appreciate all that is around us. This amounts to asking what purpose such a large universe serves. There are several possible answers to that question, and they all center upon man. Isaiah 45:18 tells us that God specially prepared the earth for man's habitation. Discovery of extrasolar planets has demonstrated how special the earth is. If the universe were very small so that there were very few planets, this would not be that compelling. It is much easier to see the special status of the earth in a huge universe with many planets.

Arguing that space has been wasted is arguing from man's perspective. Is it not a bit presumptuous to dictate what God's economy ought to be? Isaiah 55:8–9 says,

> "For my thoughts are not your thoughts,
>   neither are your ways my ways," declares the LORD.
> "For as the heavens are higher than the earth,
>   so are my ways higher than your ways
>   and my thoughts than your thoughts."

Now, the context here is the unsearchable depth of God's compassion and grace for man. However, this inconceivable depth is true of all of God's attributes. It is interesting that here God compares His mercy with the size of the universe, the very thing that we are talking about. Only if the universe is huge does this comparison make sense. Therefore, another reason why God may have made the world as large it is would be to help us understand more about Him.

Man is severely limited in his scope. For much of the 19th and the beginning of the 20th century, most astronomers thought that the universe consisted of our Milky Way galaxy and nothing more. The Milky Way is a round, flat distribution of billions of stars approximately 100,000 light years across, with the sun and its planets situated about halfway out from the center. Presumably, beyond the Milky Way was a vast expanse of emptiness. As early as the mid-18th century, a few visionaries had suggested the island universe theory, that beyond the Milky Way there were many other "universes" (today we say galaxies) similar to our Milky Way. In 1924 Edwin Hubble demonstrated that the island universe theory was true. Of course, this revolutionized our way of thinking about the universe. But even without this modern knowledge, the cosmos appears to be large. The immensity of the universe may have been part of what David had in mind when he wrote Psalm 19:1 that tells us that the heavens declare God's glory and that He is the Creator. The Apostle Paul built upon this in Romans 1:19–20, where he stated that the world around us proclaims there is a Creator so that men are without excuse. The relatively newly-discovered fact that the universe is far larger than we ever could have conceived ought to lead us to belief in a Creator, for only a powerful God could accomplish this. It is ironic that man's response since this discovery has been precisely the opposite.

Unidentified Flying Objects

Many people think that UFOs prove there is life on other planets and that beings from other planets frequently visit earth. However, this confuses UFOs with flying saucers. UFO stands for unidentified flying object, so a UFO is any object in the sky that we have not yet identified. Once an object in the sky has been identified, it no longer is a UFO; it now is an IFO, or identified flying object. On the other hand, a flying saucer presumably is a space ship from another planet. Therefore, if an object has been identified as a flying saucer, by definition it cannot be a UFO.

The vast majority of UFO reports can be explained by misidentification of known natural and man-made phenomena, converting most UFOs into IFOs. A small percentage of UFO sightings remain that defy explanation. However, that does not mean they are flying saucers; it merely means we have failed to explain them by other means. Actually, there never have been any confirmed flying saucer sightings. Hence, the evidence for alien life from UFOs or flying saucers is nonexistent.

As troubling as UFO reports might be, far more troubling are the bizarre reports of alien abductions. Some people have claimed that they were whisked away by aliens into an alien spacecraft. These claimed abductions usually are at night, while the subjects are asleep. Many claim multiple abductions. The alleged experiences during abduction vary, but there are several common themes. Some people say that the aliens showed them wonderful things, such as being taken on flights aboard the spaceships. Others say that they were given instructions or messages to mankind. These messages generally are admonishments for man to embrace better moral values. Implicit in this is universalism, the belief that all religions are equal. Some people claim that their bodies were invaded, as if they were undergoing physical examinations. Many of these body probes centered on harvesting reproductive tissues for breeding purposes. Others were overtly sexual in nature, with the implied objective being interbreeding of humans and aliens.

Tremendously significant to these claims is the fact that no physical evidence remains. This conveniently is explained away by the supposed advanced medical knowledge that the aliens possess. Virtually no one claims clear, direct memory of these alleged events. Rather, at best they are vaguely recalled as dreams. Other memories are supposedly dredged up through repressed memory therapy of various types. If memories are not clear but instead appear dream-like, perhaps they are just that—dreams.

There is a more chilling possibility. Many Christians have developed theories with regard to unexplained UFO phenomena and claimed alien abductions. Noting that 2 Corinthians 11:14 tells us that Satan masquerades as an angel of light, could demonic activity explain some of these things? There could be several possible reasons for this involvement. One could simply be to stir up interest and belief in extraterrestrial aliens. If intelligent life, similar to us, exists on other planets, then that suggests that life has arisen elsewhere in the universe. If life arises wherever conditions are conducive for life, then there is no need for a Creator. If God does not exist, then there is no such thing as sin. If sin does not exist, then men are not sinners. If we are not sinners, we are not in need of salvation. In other words, if demonic activity promotes belief in flying saucers, then it subtly undercuts belief in biblical Christianity. That is, some of the lights in the sky that people see as flying saucers could be demonic manifestations.

As for claimed encounters with aliens, some people who have made these claims seem to be sincere in their belief that these things really happened to them. Whether these are dreams or delusions, what might be the cause? Some Christians have suggested these people have come under demonic influence. Accompanying this is the suggestion that alien abductions could be invoked to explain remarkable things that many Christians think will transpire in the end times. For instance, many Christians believe that there will be a rapture of the saints prior to a seven-year tribulation. Alien abductions would make a good explanation after the fact. Of course, we must keep in mind that not all Christians believe in this particular eschatology, so we must be careful in discussing this possibility.

At any rate, belief in extraterrestrials distracts people from the true and living God. One does not have to appeal to the somewhat sensational possibility that demons are interacting with the world in this way. Just stirring up in the hearts and minds of men belief in space aliens is enough to achieve the adversary's purpose.

PART 4

Astronomy and Distortions of the Bible: Misconceptions of What Scripture Teaches about Astronomy

The final part of this book is different from the previous three parts. The book has progressed from very affirmative, reasonably clear statements about astronomy in the Bible, to less certain things about astronomy, to questions related to astronomy that the Bible does not directly address, but which can be answered using biblical principles. Now we come to some false claims about what the Bible teaches. There are many examples of things that people think are in the Bible, but are not. One of the best examples is, "God helps those who help themselves." This often is attributed to Benjamin Franklin (1705–1790), but the statement in this form actually goes back to Algernon Sidney (1623–1683). However, quotes reflecting this idea go back even further, at least to ancient Greek literature. There are many other examples of quotes popularly thought to come from the Bible but don't, such as "Spare the rod and spoil the child," "God works in mysterious ways," and "Cleanliness is next to godliness."

Just as there are these phantom biblical statements, there are ideas that many people think the Bible teaches when it actually doesn't. Chapter 13 discusses two of them. One of these ideas is the notion that the earth is flat. In the United States, most of us grew up learning a "cultural mythology" about Christopher Columbus. From childhood, we were told that until five centuries ago, everyone thought that the earth was flat and that Columbus somehow proved them wrong. According to this myth, it supposedly was the Bible and the church that taught the earth was flat, and this stifled scientific progress. In reality, people have known that the earth was spherical for more than 2,000 years. Nor did the Bible play any role in this, because the Bible doesn't teach a flat earth. As it turns out, this myth was created by atheists in the late nineteenth century to slander Christianity and promote evolution.

Unfortunately, the other issue discussed in Chapter 13, geocentrism, is not so clear-cut. While the Bible does not teach that the earth is the center of the universe, ancient pagan Greek scientists did. During the Middle Ages, the Roman Catholic Church came to embrace ancient Greek ideas, including geocentrism, as truth and wedded them to its teachings. When some scientists began to challenge geocentrism four centuries ago, the Roman Catholic Church eventually attempted to rein them in, but not for the reasons generally thought. Furthermore, this opposition from the Roman Catholic Church rapidly waned, and it hardly has been an issue since. Unfortunately, in their zeal, a few creationists today have decided that Christianity erred four centuries ago when it abandoned geocentrism. These modern-day geocentrists have been very vocal in attempting to spread their message. They are very sincere, but they are sincerely wrong.

Finally, I conclude Part 4 with three chapters devoted to the "gospel in the stars" theory. I recognize that this seems an exceptionally lengthy treatment, but it requires extensive discussion. I also ought to add that I have previously published much of what I have written here, though this version has been heavily edited. This idea is not much more than a century-and-a-half old, but it has acquired the trappings of some ancient doctrine. I have studied this subject for years, and I am constantly amazed by how poorly founded it is. Yet the notion remains stubbornly persistent among many Christians today. This baffles me. I know that many people cherish the gospel in the stars and that they find encouragement in it. However, any encouragement based upon untruth is at best false encouragement. Our commitment ought to be to truth. I ask those who believe the gospel in the stars prayerfully to consider what I have written.

CHAPTER 13

The Errors of Flat-Earth Cosmology and Geocentrism

It has become fashionable, even in otherwise theologically conservative circles, to read the Old Testament through the lens of ancient Near Eastern literature. Since the Old Testament was written in the ancient Near East, this seems to be appropriate. After all, it is reasoned, it is important to interpret Scripture within its historical context. This is true, for there are many customs and practices alluded to in the Old Testament that are difficult for the 21st century Western mind to comprehend without explanation. Of interest to creationists in the discussion of the Bible in the context of the ancient Near East are the first few chapters of Genesis. The intent of much of this work is to demonstrate that the ancient Hebrews borrowed the myths of Creation and the Flood from their neighbors. Some would argue that in borrowing these myths, the ancient Hebrews retold them in the form of a parallel story (or instead, some suggest, as an ahistorical polemic against the pagan gods worshiped by the nations surrounding Israel).[1] This approach holds that we need not view the first few chapters of Genesis as history, and thus it frees us to accept much of the secular history of the world that is widely believed today while at the same time retaining the authority of Scripture.

Integral to this assumption is that the Old Testament, and especially Genesis, reflects the cosmology of the ancient Near East. Two salient points of this cosmology are that the universe is geocentric and the earth is flat. In this chapter, we shall explore these claims, and we shall see that they are baseless. It is disheartening that some who believe in the authority of Scripture have capitulated on these points.

---

[1] The Old Testament does in fact contain a number of polemical texts, either attacking pagan deities or promoting Yahweh as the supreme, unopposed Sovereign of the universe. However, all of the Old Testament's polemics are rooted in history, not fable.

The Flat Earth

The earth certainly looks flat. We all have been taught since our earliest youth that the earth is spherical (we ought to say spherical rather than round, because something can be both flat and round, like a pizza pan). However, other than being told, how do most people know that the earth is spherical? They don't. One could argue that photographs from space show the earth to be spherical, but such photos can be faked. Furthermore, those photos have been available only over the past half century, but belief in the spherical earth goes back much further than that. How did people in the past know that the earth was spherical, and how far back did they realize this? The answer may surprise you.

The earliest recorded teaching of a spherical earth is that of Pythagoras, in the sixth century BC. How did he figure this out? A lunar eclipse occurs when the moon is exactly opposite the sun in our sky so that the earth's shadow falls on the moon. This can happen only once a month during full moon. However, there is not a lunar eclipse every full moon. This is because the moon's orbit is tilted a little more than 5° to the earth's orbit around the sun, so during most full moons, the earth's shadow passes either above or below the moon's location, thus missing the moon. The earth's shadow is larger than the moon, so we cannot see the entire shadow. But as the earth's shadow creeps across the lunar surface, its shape obviously is an arc representing a portion of a circle. Therefore, the earth's shadow is a circle. If the earth were flat and round, it could cast a circular shadow on the moon during eclipse, but only at midnight. If a lunar eclipse were at sunrise or sunset, its shadow would have some other shape, such as an ellipse, rectangle, or line. However, the earth's shadow always is observed to be a circle, regardless of the time when an eclipse occurs. A sphere is the only shape that always casts a circular shadow regardless of its orientation.

There are other proofs of the earth's spherical shape. Ancient people traveled extensively, such as by ship in the Mediterranean Sea. When traveling northward, people noticed that stars in the northern sky rose higher while stars in the southern sky fell lower. When traveling southward, the reverse happened, with stars in the southern sky rising higher while stars in the northern sky lost altitude. This can happen only if we move along an arc while traveling in the north-south direction. A similar thing happens while travelling east or west, though that is more subtle. With today's rapid transportation and accurate watches, it is easy to see, as anyone who has traveled through several time zones by airplane can attest. The sun rises and sets about three hours later on the west coast of the United States than it does on the east coast. People in the ancient world did not have rapid transportation or watches, so they could not see this as we can today, but there was another way. Ancient people noticed that a lunar eclipse did not happen at the same time for observers separated in the east-west direction. An observer in the east might see that a lunar eclipse began after sunset with the moon already in the sky. However, an observer some distance to the west might see that the eclipse was already underway when the moon rose at sunset. This can happen only if the earth is curved in the east-west direction. If the earth is curved in both the north-south or east-west directions, it clearly cannot be flat. If the earth is curved in both directions, its most likely shape is a sphere.

There is other evidence that the earth is spherical. Sometimes the claim is made that a person sees the hull of a departing ship disappear before the sail does. This would not happen on a flat earth, but it would if the earth is spherical. However, this is not clearly seen because of the distance required. Over three miles, the drop is only one foot. A ship's hull typically is much higher than one foot, so the ship must be much farther away than three miles for the hull to disappear. This would be difficult to see with the naked eye, and the telescope was not invented until four centuries ago. The concept is correct, though it probably was not practical in the ancient world. However, the concept can be demonstrated without optical aid if one attempts to look at an island, say 20 miles away. A person standing on a shore on a clear day would not see the island, unless the island had some high hills. However, if there is a bluff overlooking the shore, a person standing on the bluff can see the island. Similarly, a person stationed near the top of a ship's mast can see other ships and land before those standing on the ship's deck. This is why sailing ships often had a crow's nest near the top of a mast. Sailors in the ancient world certainly must have been aware of this effect.

The earth's spherical shape was widely known by the time of Aristotle in the fourth century BC. In the third century BC, Eratosthenes measured the earth's circumference. Eratosthenes was a Greek who lived in North Africa. Besides his role in accomplishing this feat, Eratosthenes also is known as the father of geography. Eventually he became the head librarian at the Great Library in Alexandria. Eratosthenes knew, probably by personal experience, that at noon on the summer solstice the sun was directly overhead at the ancient Egyptian city of Swenet (this is the ancient Egyptian name; the ancient Greek name was Syene. The name today is Aswan). In Alexandria, Eratosthenes constructed a gnomon, a vertical pole of known height used for measuring the length of its shadow. He measured the gnomon's length at noon on the summer solstice one year, and, using trigonometry, found that the angle between the sun and vertical was one-fiftieth of a circle. This meant that the earth's circumference was 50 times the distance between Alexandria and Swenet. From the extensive maps available to Eratosthenes, he knew the distance between the two cities and hence he computed the size of the earth.

Many people are astonished to learn these facts, because our cultural mythology is that nearly everyone in the world thought that the earth was flat until the first voyage to the New World by Christopher Columbus five centuries ago. According to this myth, a few visionaries, such as Columbus, knew better and battled against the forces of ignorance to show that the earth was spherical rather than flat. Besides being historically inaccurate, there are other problems with this commonly believed story. For instance, Columbus did not sail around the world (that was done three decades later by the crew of Ferdinand Magellan), but rather sailed to the West Indies and then back to Spain. How this accomplishment demonstrated that the earth was spherical is never explained.

According to the Columbus mythology, there was a lengthy dispute between Columbus and his critics. There was such a dispute, but it was over the size of the earth, not over the shape of the earth. None of his critics doubted that the earth was spherical and hence that one could get to Eastern Asia by sailing westward from Western Europe rather than the traditional eastward route. The question was why one would sail westward rather than eastward, given the greater distance involved. A quick glance at a globe reveals that it is indeed farther to reach Eastern Asia from Western Europe by going westward than by traveling eastward. Therefore, Columbus' critics were correct. Compounding this was that a voyage westward would take one over uncharted waters with little prospect of land along the way. At that time, the tiny ships in use were never sailed more than a few days from the sight of land. A voyage lasting several months over open water with no known landfall was extremely treacherous.

Columbus responded that the distance was not as great as most people thought. He reached this conclusion by reducing Eratosthenes' measurement of the earth's size and by inflating the estimated distances from Western Europe to Eastern Asia going eastward. Columbus was influenced by reports of lands to the west, which he assumed were Asia. Five centuries earlier, Vikings had explored and briefly settled in Newfoundland. The Vikings tried to keep this information to themselves, but such a thing was not possible. According to Irish legend, Brendan of Clonfert, a sixth-century Irish monk, sailed from Ireland to a land to the west and returned to Ireland. Some maps at the time of Columbus depicted Saint Brendan's Island in the North Atlantic somewhere west of Northern Africa.

How did this nonsense about the flat earth and Columbus get started? There are three principle culprits in this case. The first was Washington Irving, who in 1828 published *A History of the Life and Voyages of Christopher Columbus*, a supposed historical account of the life of Columbus. In reality, it more closely resembled a modern movie adaptation of real events mingled with imagined ones. At the time, the genre of historical fiction did not yet exist, so many readers understandably assumed that it was historically accurate. This book appears to be the primary source of the myth of the widespread belief in the flat earth. The second culprit was John William Draper, who published *History of the Conflict between Religion and Science* in 1874. Draper's book largely was anti-Roman Catholic in that it singled out the Roman Catholic church as the institution that opposed intellectual progress during the Middle Ages. The third culprit was Andrew Dickson White, who published *A History of the Warfare of Science with Theology in Christendom*. In his book, White coined the term *conflict thesis*, the belief that religion, and more specifically Christianity, held back the progress of science. Both Draper and White were atheists, so they had a vested interest in criticizing belief in God. While less read than Draper's book, White's book proved to be more significant, partly because it appeared less anti-Roman Catholic, but also because it appeared to be better documented. The conflict thesis enjoyed widespread support for much of the 20th century, though its stock has fallen tremendously among historians of science over the past half century. Despite this, the conflict thesis remains very popular in the minds of the public.

The conflict thesis drew heavily upon the Galileo affair, which we will discuss shortly. It also drew heavily upon the church's negative reaction to Darwinism, which was a current affair in the late 19th century. The conflict thesis also relied upon the supposed insistence on the part of the church at the time of Columbus that the earth was flat. The message was very clear—the church had gravely erred on the matter of the shape of the earth, so the church had the opportunity to redeem itself by getting in on the ground floor of Darwinism. In many respects, this ploy worked, because many Christian leaders and denominations capitulated to arguments for biological evolution and a millions of years old earth. We are saddled with this legacy today after more than a century.

The other enduring legacy is, of course, our cultural mythology about the supposed widespread belief in the flat earth. In response, in 1991 the medieval scholar Jeffrey Burton Russell published *Inventing the Flat Earth: Columbus and Modern Historians*. In his book, Russell set the record straight on the history of man's understanding of the shape of the earth. Since ancient times, in the West, educated people knew that the earth was spherical. The church never taught that the earth was flat. Despite the truth, the myth of the flat earth persists.

Notice that in our discussion thus far, we have not discussed biblical passages. That is because there are not any biblical texts that teach that the earth is flat. However, skeptics have chosen to interpret a few verses as endorsing a flat earth. For instance, the phrase "ends of the earth" appears 28 times in the King James Version (e.g., Job: 28:24; Isa. 41:9; Jer. 16:19; etc.), though more modern translations may vary on the wording in some of the occurrences. Skeptics insist that this must refer to a physical edge to the earth, which could exist only if the earth were flat; ergo, the Bible teaches a flat earth. Actually, the phrase "ends of the earth" is idiomatic, and it refers to the remotest parts of the earth. In some contexts, it may allude to the four compass directions. The phrase "four corners of the earth" appears three times in the Bible. Surely, the skeptics claim, this must refer to a flat, square earth—thus proving that the Bible teaches a flat earth. There are some examples of flat earth cosmologies from the ancient world, but they always consisted of a flat, round earth. A circle was considered a much more perfect shape than a square, so none of the flat earth cosmologies were square. If a square flat earth were the cosmology of the Bible, then it would have been at odds with every other ancient flat earth cosmology. Therefore, this attempt by the skeptics to claim that the Bible teaches a flat earth does not square with the facts of history. Nor does it square with the meaning of the passages in question. The one occurrence in the Old Testament, Isaiah 11:12, should be taken in the same manner as other Old Testament passages mentioning the "ends of the earth." It is likewise an idiomatic expression. The two New Testament occurrences of "the four corners of the earth" are in the book of Revelation. Revelation 7:1 speaks of four angels standing on the four corners of the earth and restraining the four winds of the earth. The four winds obviously refer to the four directions of the wind: north, south, east, and west. This repetition ("four angels...four corners...four winds") makes it clear that this is an idiom referring to the four compass directions. The phrase "four corners of the earth" from Revelation 20:7 also is idiomatic, referring to the four directions.

Many Christians try to respond to these claims of skeptics by saying that the circle of the earth in Isaiah 40:22 refers to a spherical earth. However, this verse is ambiguous. It might refer to a spherical earth, or it may not. It may refer to the circular horizon that we can see on the earth. Indeed, some skeptics respond by saying that God above the earth views the earth as round and flat, demonstrating that Isaiah 40:22 teaches a flat earth. (Notably, some of these same skeptics claim that the Bible teaches a square flat earth, which places them in an inconsistent position.) Unfortunately, many Christians have tried to use Isaiah 40:22 to prove that the Bible is inspired by claiming that this verse reveals knowledge that generally was not known at the time it was written. However, this in itself buys into the flat earth myth. While Isaiah, written in the eighth century BC, predates the earliest recorded mention of a spherical earth by two centuries, it is possible that the earliest *surviving* record of a spherical earth was not the first recognition of the earth's true shape. That is, some people as early as the eighth century BC may have known (apart from special revelation) that the earth is spherical.

One other Bible verse has been put forth as an anti-flat earth statement. Job 26:7 says that God "hangs the earth on nothing." Ancient flat earth cosmologies required that the flat earth rest upon something, be it water, the back of a turtle, or the backs of elephants. Though this verse does not explicitly or even implicitly address the earth's shape, it does contradict all of the ancient flat earth cosmologies that we know of. Moreover, it conforms to what is known from modern observational science. At any rate, the Bible does not teach that the earth is flat, nor does it clearly teach that the earth is spherical. The Bible contains no passage that addresses the shape of the earth.

Geocentrism

The question of geocentrism is not so clear. Like the earth's shape, the Bible does not address the question of whether the earth orbits the sun or the earth is the center of the universe. For most of history, people believed the geocentric theory, the idea that the earth remained motionless as the sun, moon, and planets orbited the earth. Many people erroneously think that this was for philosophical or theological reasons. Those reasons came later, but the original reason for belief in the geocentric theory was the lack of stellar parallax. Ancient astronomers considered the heliocentric theory, the belief that the earth was just one of several planets that orbited the sun. However, if the heliocentric theory were true, we would expect that the stars would shift position back and forth annually as we orbited the sun. Ancient astronomers looked for stellar parallax, but did not see it. Therefore, being good scientists, they rejected the heliocentric theory. This left the geocentric theory as the only viable alternative.

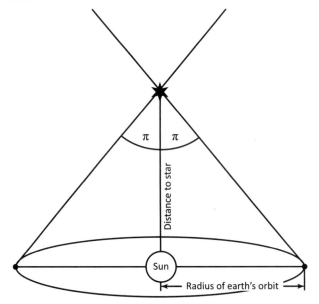

The earliest record that we have of someone advocating the heliocentric theory was Aristarchus in the third century BC. He based his conclusion upon his measurements of the size and distance of the moon and sun. He found that the diameter of the sun was about seven times the diameter of the earth. However, in terms of volume, the sun must have been hundreds of times greater than the earth. He thought it more reasonable that the smaller body should orbit the larger body. Actually, his size measurements were far off, because the sun's diameter is more than 100 times that of the earth's diameter.

Aristarchus was well aware of the lack of stellar parallax, so this required explanation. Aristarchus claimed that the stars were much too far away for us to readily see parallax. As it turns out, Aristarchus was correct about this. The amount of stellar parallax depends upon distance, with the closest stars having the greatest parallax. The closest star has a parallax of less than a second of arc, the apparent diameter of a dime viewed at a distance of three miles! The first parallax measurements were not until the 1830s, more than two centuries after the invention of the telescope.

There are two basic forms of geocentrism. What we have described thus far refers to the observed motion due to the earth's revolution around the sun each year. This is what most people think of in terms of geocentrism—that the earth does not revolve but rather it remains stationary as the sun and other bodies orbit it. However, the earth rotates on its axis each day. This rotation causes the sun, moon, planets, and most stars to rise in the east, move westward across the sky, and set in the west. Alternately, one could say that the earth does not rotate but that the entire cosmos spins around the earth each day. This is the other basic form of geocentrism, that the earth neither revolves nor rotates. The earliest mention of the earth's rotation being responsible for daily motion is from Heraclides in the fourth century BC. Ancient cosmologies were more split on the question of whether the earth spun or if the entire universe spun each day. However, a non-rotating earth with a spinning cosmos was the dominant view.

In time, other arguments for the geocentric theory became attached to this view. One argument was that as we feel other motion, such as riding a horse or wagon, we ought to feel motion as the earth moved. Since we do not feel such motion, the earth must be stationary. Of course, this overlooked the causes of the moving sensation, such as bumps and wind resistance. Another objection to the earth's motion was the belief that the moon in its obvious monthly orbit of the earth would be left behind if the earth were moving.

A philosophical or theological objection was that being on earth, man ought to be at the center of the universe. Today this generally is thought of as being in the context of the centrality of man within Christianity. That is, because man is the center of God's attention, man must be at the physical center of the universe. However, this belief predated the Christian era, having derived from an ancient pagan Greek idea based upon a faulty view of gravity. To explain why objects fell downward, the ancient Greeks posited that the normal direction of the movement of things in the physical world was downward. Objects in the heavens, such as the sun, moon, and stars, did not share in this generally downward tendency, because they were perfect. The higher that a body was in this hierarchy, the more perfect it was. This perfection enabled most heavenly bodies to avoid falling to earth and could remain in motion indefinitely. This dichotomy between the terrestrial and heavenly persisted until the development of science as we know it in the 17th century.

The underlying principle of downward movement unified both physical and moral phenomena in the terrestrial world in a form of equivocation. Objects fell downward because that was what they did. Likewise, morality fell downward because that is what it did. The underworld, where the souls of those who died departed to, was below the ground. The underworld included Tartarus, which is roughly equivalent to hell (cf. Luke 16:23). Presumably, lava and other hot emissions from inside the earth were material from Tartarus. Thus, man on the earth's surface was perched precariously just above perdition. Hence, man was not in a favored location, but was in precisely the opposite of a favored location. This concept has been flipped completely around in the modern view.

The Ptolemaic Model

As the planets revolve around the sun in their respective orbits, the motion that we observe from our moving platform as the earth revolves around the sun is very complicated. The distant stars remained fixed with respect to one another, though the stars we see each season change because of our revolution, and the stars we see throughout the night change due to our rotation. Both the moon and the sun move eastward with respect to the stars. Each day the moon averages a little more than 13° in its motion to complete one circuit per month, and the sun moves nearly a degree per day to complete one circuit per year. The five naked-eye planets, Mercury, Venus, Mars, Jupiter, and Saturn, appear as bright stars that usually travel eastward through the stars. However, from time to time the planets reverse direction, traveling west to east with respect to the stars before resuming their eastward motion. We call this apparent backward motion *retrograde motion*; the normal eastward motion is prograde, or direct motion. During retrograde motion the planets bob up and down slightly so that retrograde motion is a broad loop or a stretched out S-shape.

The cause of retrograde motion is easy to see in the heliocentric theory. As the moon orbits the earth, it moves eastward in the sky. As we orbit the sun, the sun appears to us to move eastward through the stars. The normal eastward motion of planets is due to their orbital motion around the sun. Superior planets (Mars, Jupiter, and Saturn) are planets with orbits larger than earth's orbit. Superior planets not only have larger orbits than the earth, but they move more slowly than the earth does. Therefore, from time to time the earth overtakes and passes a superior planet. As the earth overtakes a superior planet, that planet appears to move backwards (westward), similarly to how a car that we pass on a highway appears to move backwards. This is retrograde motion. Once the earth has moved sufficiently forward in its orbit, the superior planet resumes its normal prograde motion. Inferior planets (Mercury and Venus) are those planets with orbits smaller than earth's orbit. Inferior planets move more quickly than the earth does, so they occasionally overtake the earth in its orbit. This is how inferior planets exhibit retrograde motion. The bobbing motion is due to the fact that each planet has a slightly different orbital plane so that each one makes a small angle with respect to the earth's orbital plane.

Heliocentric Model of Retrograde Motion

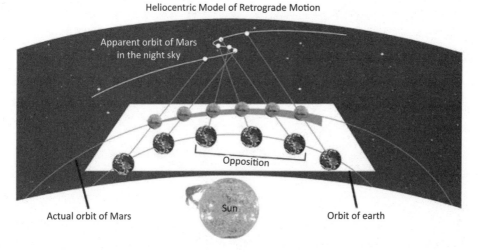

How can one explain retrograde motion in a geocentric model, where the earth is the center of motion and hence cannot pass other planets nor be passed? One could propose that retrograde motion is just the erratic nature of planets. However, this does not seem satisfying for objects in the heavenly realm, where all is supposed to be perfect. The ancient Greeks considered the circle to be the perfect shape, so it seemed reasonable that the planets, being in the perfect heavenly realm, ought to move on circular paths. Furthermore, the ancient Greeks thought that only uniform motion along their circles was appropriate for such perfect bodies. With these philosophical/theological constraints, it was not possible to explain retrograde motion simply. However, in the mid-second century AD Claudius Ptolemy, an astronomer at the Great Library in Alexandria, Egypt developed a theory that could explain retrograde motion. There were many notable people who shared his last name, but because of the prominence due to his model, we normally reserve use of the name Ptolemy for him.

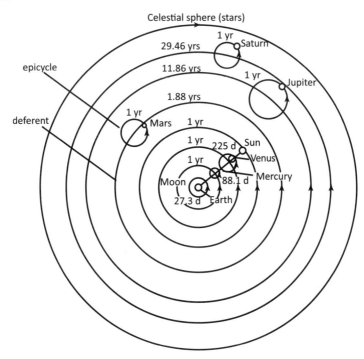

How did the Ptolemaic model work? Ptolemy suggested that each planet moved at a uniform rate on a circle that he called an epicycle. The center of each epicycle in turn moved at a uniform rate on a larger circle that was centered on the earth. These larger circles were called deferents. Both the motion of the planets on their epicycles and their epicycles on their deferents were in the same direction, counterclockwise as viewed from above the earth's Northern Hemisphere. The observed motion of a planet as seen from the earth would be the combined motion of the planet on its epicycle and the epicycle's motion on its deferent. Most of the time, this combined motion would be in the same direction. However, a planet occasionally would travel on the portion of its epicycle that is closest to earth. During this time, the planet's motion on its epicycle opposed the motion of the epicycle on the deferent, producing retrograde motion. By adjusting the sizes of the epicycle and deferent, and the rate at which a planet and epicycle moved, Ptolemy was able to match the observed retrograde motion.

While the introduction of epicycles qualitatively described the retrograde motion of planets, the model required further modifications to produce a perfect match. Since the orbits of the other planets are slightly inclined to the earth's orbit, they bob up and down. Ptolemy explained this by adding a second smaller epicycle for each planet that was oriented perpendicular to the larger epicycle. Furthermore, planetary orbits are not circles, but rather they are ellipses with the sun at one focus of each ellipse. The deviation of each orbit from a circle is not much, so Ptolemy handled this by displacing the earth slightly from the center of the circle. Technically, this is a departure from a truly geocentric model, but none of his followers seemed to care. Finally, the planets do not move along their elliptical orbits uniformly, but instead move most quickly when closest to the sun (perihelion) and most slowly when farthest from the sun (aphelion). Ptolemy overcame this difficulty by having the epicycles move at a uniform rate with respect to a point called the equant—not the centers of their respective deferents. The equant of each deferent was collinear with the earth and the deferent's center, with the center exactly halfway between the earth and the equant. This too was a departure from the constraints of the model (uniform motion on circles); but, again, Ptolemy's followers did not seem to care.

In addition, the sun and moon moved along deferents in the Ptolemaic model. Because the orbits of both the earth and the moon are ellipses and neither of them moves at a uniform rate along their ellipses, Ptolemy shifted the earth from the centers of the moon's and sun's deferents, and he introduced equants for them. The deferent of each body ranged in size in order of increasing distance from the earth as understood in ancient Greek cosmology. That order was determined by the rate at which objects appeared to move with respect to the stars, with the most rapidly moving objects closer to earth. The order, in terms of increasing distance, was the moon, Mercury, Venus, the sun, Mars, Jupiter, Saturn, and then the stars (which did not move).

Because the Ptolemaic model was able to explain the motion of the planets so well, it was an immediate success. Ptolemy included his model in his 13-volume *Mathematical Syntaxis*. The popularity of the Ptolemaic model resulted in the endless copying of this book and thus its preservation. This is fortunate for us, because in his book Ptolemy described his knowledge of ancient Greek astronomy. Without Ptolemy's work, we would not know much about the history of ancient Greek astronomy, because Ptolemy discussed the works of his predecessors, whose works do not survive. During the medieval period, Ptolemy's work was largely lost in the West, but it remained in the eastern portions of what had been the Roman Empire. After the Muslim conquest, Arabic astronomers perpetuated Ptolemy's work. They were so impressed with it that they referred to it as "The Greatest," or *Almagest*. We generally refer to Ptolemy's work by that Arabic title today. Many late medieval copies of the *Almagest* are extant.

The Ptolemaic model was considered scientific truth until the 17th century. In terms of longevity—15 centuries—it is the most successful scientific theory of all time. No other theory comes even close to matching it. The strength of the Ptolemaic model was that it could be modified slightly to account for any discrepancies that arose between its predictions and observations. And over time, people did add to the model. For instance, the original Ptolemaic model required at least 11 epicycles, two for each planet and one for the moon. However, since the Ptolemaic model did not properly model reality (using circles rather than ellipses, for example), discrepancies between its predictions and reality slowly accumulated. As those discrepancies mounted, the fix was to add more epicycles. By the end of the 16th century, some modifications of the Ptolemaic model required over 100 epicycles. Eventually, people began to realize that the strength of the Ptolemaic model, the fact that it could be modified at will, also was its undoing. As the Ptolemaic model became very complicated, people realized that reality probably was simpler.[2] This caused various scientists to reconsider the heliocentric model. Credit for adoption of the heliocentric model normally goes to Nicholas Copernicus (1473–1543). Many people erroneously think that Copernicus originated the heliocentric theory. He did not, because some ancient Greeks believed the heliocentric theory. However, Copernicus did make several important contributions. Just before his death, Copernicus published *De Revolutionibus Orbium Coelestium*

---

[2] This principle, often called the law of parsimony or Ockham's razor, for William of Ockham (1287–1347), is that when confronted with two explanations for something, all other things being equal, the simpler explanation usually is the correct one.

(*On the Revolutions of Heavenly Spheres*), in which he gave arguments for the heliocentric model. Furthermore, in his book Copernicus derived the true orbital periods and relative orbit sizes for the planets. We have no record that this had ever been done before. His book attracted much attention right away and for nearly a century more.

Between the introduction of the Ptolemaic model and the 17th century, two important philosophical beliefs about the Ptolemaic model developed. First, people began to equate the Ptolemaic model with reality. Ptolemy had not meant his model as a statement of reality. Rather, he used the best mathematics of his time, geometry, to express his model for computing planetary orbits. Today we use algebra to compute planetary positions, but no one would confuse the algebraic expressions used for reality. However, somewhere along the line, people began to view Ptolemy's method of planetary position computation as how the world actually worked. Second, largely through the work of Thomas Aquinas (1225–1274), the Roman Catholic church came to embrace the Ptolemaic model. Aquinas was quite impressed with the work of Aristotle (384–322 BC), and so Aquinas attempted a synthesis of Roman Catholic teaching and Aristotelian philosophy. Prior to Aquinas, the Roman Catholic church had embraced the teachings of Plato (late fifth and early fourth century BC) because of the influence of Augustine (354–430 BC). While Plato concentrated on philosophy, Aristotle concentrated on both philosophy *and* science. Ptolemy was very much in the mold of Aristotelian thinking, so it was natural for the Roman Catholic church to incorporate the Ptolemaic model into its dogma as well. Through Aquinas' teaching, the Roman Catholic church gradually came to rely more upon science and other non-biblical sources in determining truth.

This shift in understanding truth is underscored by the Galileo affair, though many people miss this point. Galileo Galilei (1564–1642) was an Italian mathematician and astronomer. People often incorrectly credit Galileo with the invention of the telescope. Though he did not invent the telescope, he probably was the first person to put the telescope to use studying astronomy. With his telescope, Galileo saw that Venus went through complete phases, which could happen only if Venus orbited the sun. This clearly challenged the Ptolemaic model. He also discovered four satellites, or moons, orbiting Jupiter. This challenged the Ptolemaic model, because here were four objects that did not orbit the earth. Furthermore, it refuted one of the arguments against the heliocentric theory, that the earth would leave the moon behind in its motion, because Jupiter obviously moved, yet its satellites remained in orbit around it. Moreover, Galileo saw spots on the sun and craters on the moon. This contradicted the ancient Greek concept of the perfection of the heavenly bodies, with which Aristotelian philosophy agreed.

However, much earlier, Galileo had read Copernicus' book and had come to believe the heliocentric theory. Even before Galileo was born, Copernicus' book had sparked much discussion among scientists, but it had failed to attract the attention of many theologians. Theologians were not very concerned with the heliocentric theory, but scientists of the day were very interested in it, because it directly challenged the foundations of their beliefs. In 1610, Galileo published *Sidereus Nuncius* (*Starry Messenger*), in which he described his telescopic observations and advocated the heliocentric theory. This book was very controversial, and soon other scientists complained to Roman Catholic officials, and the Catholic officials looked into the matter. Galileo asked that the church stay out of this scientific squabble, but to no avail. Previously, the Roman Catholic church had deemed the heliocentric theory acceptable, but merely as a mathematical and hypothetical perspective, and so it was not to be taught as truth as Galileo had done. In its response in 1616, the Roman Catholic church reaffirmed this position, but it went further. It declared heliocentrism was a matter of heresy, because it ran counter to the church's official teachings. This move eventually resulted in the banning of heliocentric material, such as Copernicus' books and the writings of Johannes Kepler (1571–1630). Note that the teaching of the Catholic church which the heliocentric theory challenged was based upon Aristotle, *not* Scripture.

The church also ordered Galileo to cease teaching the heliocentric theory as truth. He must have obeyed this order for the most part, because the next 15 years were uneventful. However, in 1632, Galileo published his *Dialogo Sopra i due Massimi Sistemi del mondo* (*Dialogue Concerning the Two Chief World Systems*). The *Dialogue*, as it is often called, was unusual in its day, because it was written in Italian rather than Latin. Galileo apparently did this to reach a broader audience. The *Dialogue* consists of a discussion between three characters, Salviati, Sagredo, and Simplicio. Salviati supported the Copernican model, while Simplicio defended the Ptolemaic model. Salviati was smart, but Simplicio decidedly was not. Galileo claimed to have named Simplicio for the Aristotelian philosopher Simplicius (490–560), but it probably is not coincidental that the Italian form of that name is roughly equivalent to "simpleton." Compounding the problem was the witty and not so neutral observer Sagredo who often sided with Salviati. Pope Urban VIII had insisted certain arguments be included in the book, which resulted in his words coming from Simplicio. Whether Galileo intentionally did this to make the Pope look bad or not, it had the effect of alienating Pope Urban VIII (1568–1644), who previously had supported Galileo. Over the years, Galileo had managed to offend many people, and in many respects, this was just the last straw. The following year he was tried and found guilty of violating the order from 17 years earlier. Technically, this made him guilty of heresy, since the heliocentric theory had been branded as such. However, unlike true heretics of his time who endured harsh punishment and even execution, Galileo was sentenced to house arrest for the rest of his life (he lived another decade and died at age 77).

Most people today take away from the Galileo affair the wrong message. It is commonly taught that Galileo was minding his business pursuing the heliocentric theory when the Roman Catholic church attempted to muzzle him because what he taught was contrary to Scripture. The lesson supposedly is that religion ought not to interfere in questions dealing with science. Often a comparison is made to today with creationists assuming the role of the Roman Catholic church and evolution being the equivalent of the heliocentric theory. However, this is not what the lesson of the Galileo affair ought to be. In reality, the Roman Catholic church did not oppose the heliocentric theory because it was contrary to Scripture, but rather because it opposed scientific teaching of that time. It was the scientists who pressured the Roman Catholic church to rein in Galileo. Galileo was challenging the scientific status quo. Therefore, if one is to make an analogy to today, then the correct analogy is to identify creationists with Galileo, because creationists are challenging the status quo of evolution. Furthermore, the Roman Catholic church wielded considerable power and authority that no church does today. Instead, that power and authority is wielded by the academic and scientific communities today, which further makes the current evolutionary establishment the equivalent of the Roman Catholic church in the Galileo affair.

The primary alleged refutation of Galileo came from Aristotle and Ptolemy, not from Scripture. This underscores the fact that the Galileo affair was a scientific squabble, not a conflict between religion and science. However, there were some Bible verses that were brought into the discussion at the time.

The most commonly used verse was Joshua 10:13, where Joshua commanded the sun and the moon to stand still. From this, people inferred that the sun and moon must be moving, not the earth. However, this confuses the question of the earth's rotation and revolution. One can have a stationary earth that rotates, as geocentrism generally is understood in the context of the earth not orbiting the sun, so this verse does not really address that issue. Furthermore, even today astronomers believe that the earth both rotates and revolves, yet they speak of sunrise, sunset, moonrise, moonset, and the (apparent) motion of the sun, moon, and stars across the sky. This is the language of observation, and in that sense the sun, moon, and stars do rise and set—while at the same time we understand that this apparent motion is due to the earth's rotation. Furthermore, teaching cosmology is not the point of this passage, but rather it is the historical account of the Battle of Gibeon. As a modern example, consider the Battle of the Philippine Sea in June 1944. Probably every account of that battle describes the significance of US naval planes returning to their carriers after sunset and low on fuel. No one would interpret such accounts as endorsing geocentrism. At any rate, those who did enlist Scripture to refute Galileo used passages that they had (wrongly) interpreted in terms of the geocentric model (please see chapter 6 for more discussion of Joshua's long day).

Modern Geocentrism

There are some Christians today who maintain that the Bible teaches geocentrism. They claim that the problem with reinterpreting the Bible in terms of evolution and billions of years did not start in the 19th century, but that the roots of the problem go back to the Galileo affair. In their estimation, the problem began with adopting the heliocentric theory, because that called into question the alleged geocentric teachings of the Bible. Which verses do they claim in support of this? Obviously, the passage just mentioned, Joshua 10:13, is one. Another is 1 Chronicles 16:30, which reads,

> Tremble before him, all the earth;
> yes, the world is established; it shall never be moved.

Surely, they reason, if this verse means anything, it means that the earth does not move. Or does it? What is the context of this verse? In 1 Chronicles 15, David brought the Ark of the Covenant to Jerusalem, and in 1 Chronicles 16:1–7, David brought the Ark into a tent and offered offerings before it. The narrative continues with a song of thanksgiving in 1 Chronicles 16:8–36, which includes the verse in question. Being a song, it amounts to poetry. Indeed, portions of this song are repeated in Psalms 96, 98, 100, and 106. For instance, the relevant portion of 1 Chronicles 16:30 appears word for word in Psalm 96:10 (and so it too is included in the geocentrists' list of verses supporting their position). However, poetry often contains imagery and figurative language. For instance, 1 Chronicles 16:31, which immediately follows 1 Chronicles 16:30, states,

> Let the heavens be glad, and let the earth rejoice,
>   and let them say among the nations, "The LORD reigns!"

Notice here that the heavens and earth are personified, for the heavens are described as being glad and the earth is commanded to rejoice. It is doubtful that even the most literal of literalists (including geocentrists) would insist that the heavens can feel emotions or that the earth can speak. They would agree that these are poetic allusions. Yet modern geocentrists insist that 1 Chronicles 16:30 must mean that the earth does not move in some literal sense. The phrase "the world is established; it shall never be moved" from 1 Chronicles 16:30 and Psalm 96:10 essentially says the same thing two different ways. In these verses, the earth not moving is not to be understood as a reference to its lack of motion, but to the fact that God created the earth to endure—God created *and* continually sustains the earth. The modern geocentrists are claiming that these verses mean something other than what the authors intended them to mean in their respective literary and historical contexts.

Another verse that the modern geocentrists claim for support is Psalm 93:1, which reads,

> The LORD reigns; he is robed in majesty;
>   the LORD is robed; he has put on strength as his belt.
> Yes, the world is established; it shall never be moved.

Notice that the relevant phrase, "the world is established; it shall never be moved," is exactly the same wording as 1 Chronicles 16:30 and Psalm 96:10 (and the Hebrew wording underlying the translation is likewise the same). However, in Psalm 93:1 the translators of the King James Version chose to render the word *establish* as "stablish." Most modern geocentrists claim that this significant, because the words *stablish* and *establish* supposedly mean different things. The argument thus goes that this alleged different meaning proves that the Bible teaches geocentrism. It does not matter that in the very next verse the King James Version translated *the same Hebrew* word as "establish." Thus, modern geocentrism relies upon the King James Version as opposed to any other translation or even the original Hebrew text.

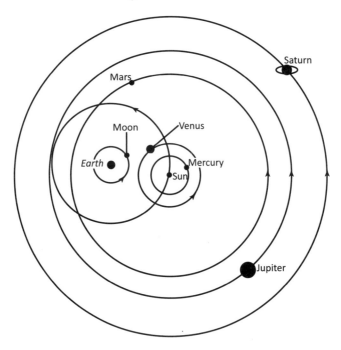

Notably, modern geocentrists do not uphold the Ptolemaic model. Rather, they support the Tychonic model. Tycho Brahe (1546–1601) proposed his model as a compromise between the geocentric and heliocentric models. In his theory, the other planets orbited the sun, and the sun in turn orbited the earth and carried the other planets along with it. In this respect, the Tychonic model eliminated the need of at least the major epicycles. Furthermore, it amounts to a coordinate transformation from the heliocentric model. A coordinate transformation best describes what modern geocentrists attempt to do. They view gravity as providing the force causing the planets to orbit the sun, just as in the heliocentric model. The difference is that space is somehow attached to the earth, so as the earth orbits the sun, it carries space along with it. If space is attached to the earth, then it is impossible for the earth to move with respect to space, so the earth remains motionless. Even though it moves. This amounts to geocentrism by definition, for the earth is at rest with respect to itself.

Modern geocentrists also prefer their own reconstruction of history. For instance, they claim that the Tychonic model was a major contender among cosmological theories four centuries ago. Admittedly, it did have its adherents for a while. Part of its supporters' motivation was that the Tychonic model could explain the phases of Venus, whereas the Ptolemaic model could not. However, as with many compromises, it did not really satisfy most people. The tendency in these sorts of disputes is to abandon an idea once it is shown to be inadequate rather than compromising with it. Later discoveries, such as James Bradley's (1693–1762) discovery of aberration of star light in 1728, the measurement of parallax in the 1830s, and the observation of annual variation of Doppler motion in the light of stars in the 19th century directly confirmed the heliocentric theory. Modern geocentrists have interpreted these findings in terms of the Tychonic model, but this amounts to altering the theory to fit the data.

In conclusion, the modern geocentric movement has a heavy dependence upon the text of the King James Version alone. Due to this, there is commonly an air of piety in the arguments of geocentric proponents. Added to this is the problem of selective use of hyper-literalism, as geocentrists today acknowledge that the Bible contains figurative language, but *they* are the arbiters of when and where figurative language exists. While modern geocentrism has made inroads with certain groups of Christians today, it is based upon a faulty view of Scripture.

CHAPTER 14

The Gospel in the Stars: Introduction and Alleged Basis

What Is the Gospel in the Stars?

Anyone even vaguely familiar with astronomy knows that we divide the sky up into groupings of stars called constellations. The word *constellation* comes from the two roots *con*, meaning with, and *stella*, meaning star. The constellations are outlines or stick figures representing various objects, people, animals, and mythological beings, such as centaurs. While some of the constellations bear resemblance to their namesakes, alas, many of them do not. Consequently, it takes quite an imagination to make out the figures of many of the constellations.

The names of most of the constellations are unfamiliar to the general public, but there are 12 constellations that are well known. These are the 12 signs of the zodiac: Aries, Taurus, Gemini, Cancer, Leo, Virgo, Libra, Scorpius, Sagittarius, Capricornus, Aquarius, and Pisces. The word *zodiac* comes from two Greek words, *zoe*, meaning animal (we get the word zoo from this root), and *diakos*, meaning circle. Hence, the word *zodiac* means "circle of animals," because eleven of the 12 zodiacal constellations represent living things.

The constellations in general probably are mentioned in Job 38:32. In the King James and Revised Standard Versions, the word *mazzaroth* appears. This is a transcription of the Hebrew מַזָּרוֹת (*mazzārôṯ*). However, the New American Standard Bible and New International Version translated this word as "constellations." The word may refer to all the constellations, or it may refer to just the zodiacal constellations. This word appears just this one time in the Bible, but a similar word, מַזָּלוֹת (*mazzālôṯ*; perhaps a variant spelling), appears once as well, in 2 Kings 23:5. The King James Version translated this word as "planets," but the Revised Standard Version, New American Standard Bible, and New International Version all render it "constellations." The Septuagint renders it μαζουρωθ, exactly the same word appearing in its translation of Job 38:32. As with *mazzārôt*, we do not know for sure if *mazzālôṯ* refers to all the constellations or just the zodiacal constellations.

The zodiac forms the basis of astrology, an ancient pagan cultic system. How does this supposedly work? The stars move, but they are so far away that even over thousands of years the shapes of the constellations do not appreciably change. The constellations that we see throughout the night change as a result of the earth's rotation. And as the earth orbits the sun, the constellations change from season to season, but they always return in due course each year.

Against this backdrop of the changeless stars, there are seven bright objects that move. The sun is the most obvious moving object. As we orbit the sun, the sun slowly appears to move through the stars, taking a year to complete one circuit. It is this motion that causes the stars and constellations that are visible throughout the year to change. We call the apparent path of the sun through the stars each year the *ecliptic*. The ecliptic passes through all 12 of the zodiacal constellations.

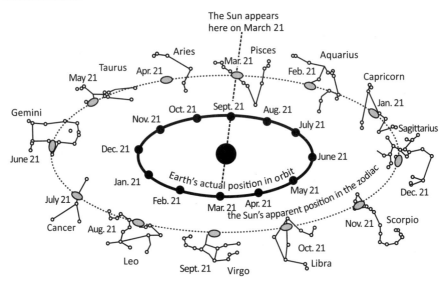

Nearly as obvious as the sun's motion is the moon's motion. The moon takes a month to orbit the earth, so over the course of a month the moon appears to complete one circle through the stars. Because the moon's orbit around the earth lies in nearly the same plane that the earth orbits the sun, the moon follows a path through the stars that lies within 6° of the ecliptic. This maximum angular distance of the moon's path on either side of the ecliptic roughly defines the zodiac.

Less obvious is the motion of the five naked-eye planets—Mercury, Venus, Mars, Jupiter, and Saturn. These planets appear as bright stars in our sky. The other two planets, Uranus and Neptune, generally are too faint to be visible to the naked eye. Venus and Jupiter are the planets which appear the brightest, even brighter than the brightest stars. The other three naked-eye planets are fainter, but they still rival the brightest stars in brightness. The orbits of the planets lie in nearly the same plane as the earth's orbit, so the planets are never very far away from the ecliptic, usually falling in the band of the zodiac defined by the moon's path. However, the planets follow very complicated back-and-forth motions with respect to the stars of the zodiac. The backward, or retrograde, motion is due to the combined orbital motions of the earth and the other planets.

Hence, the 12 zodiacal constellations are special, because only they host the sun, moon, and planets. To many of the ancients, the sun was an important deity, for it provided light and warmth, and it also caused plants to grow. The moon also was a god (or goddess), because it ruled over the night (cf. Gen. 1:16), and also because it provided light at night, caused the tides, and served to help mark the passage of time (cf. Gen. 1:14). The planets do not seem to have such noticeable effects on us, but often the ancients viewed them as deities too, as evidenced by the fact that the names for the planets that they handed down to us are the names of Roman gods.

As discussed in Chapter 5, astrology is an ancient pagan religion, and as such is foreign to the Bible and Christianity (Deut. 4:19; Isa. 47:13–14). Even apart from the 12 constellations of the zodiac, many of the remaining constellations are steeped in lore laden with pagan mythology. Given this pagan connection, how is the Christian to view the constellations? One approach is to acknowledge the pagan history, but also to realize that this history is false and as such has no power over us.

Another approach to the constellations holds that they represent the vestiges of a primal gospel presented to early man before God's written revelation. We call this view "the gospel in the stars." According to this theory, God presented the full story of His plan of salvation to Adam, and Adam's descendants through Seth's godly line preserved that knowledge. Either God ordained patterns in the sky to teach the lesson, or early men crafted the gospel in the stars to preserve the message. With the coming of the written Word of God, the gospel message in the stars was no longer needed and hence faded from use. Additionally, the passage of time allowed ungodly men to pervert the original gospel in the stars, mingling it with much pagan mythology and ultimately turning it into the religion of astrology. However, vestiges of this long-lost truth still exist, and proponents of the gospel in the stars have attempted to reconstruct this supposed gospel message. In this chapter and the two to follow, we will examine the gospel in the stars thesis.

History of the Gospel in the Stars

Supposedly, the lost gospel in the stars was rediscovered in the mid-19th century by the English woman Frances Rolleston, whose work was posthumously published in the book, *Mazzaroth; or, the Constellations* in 1865. Her book soon influenced others. An early example of her influence is seen in James Gall's 1871 *Primeval Man Unveiled: or, the Anthropology of the Bible*, which includes a chapter on "Antediluvian Theology," fully crediting Rolleston as his source. Better known and more complete treatments with embellishment are the books of the American pastor Joseph A. Seiss (*The Gospel in the Stars*, 1882), and the English theologian E. W. Bullinger (*The Witness in the Stars*, 1893). To understand Rolleston's key role in developing the gospel in the stars thesis, consider this from the preface of Bullinger's book:

> Some years ago it was my privilege to enjoy the acquaintance of Miss Frances Rolleston, of Keswick, and to carry on a correspondence with her with respect to her work, *Mazzaroth: or, the Constellations*. She was the first to create an interest in this important subject. Since then Dr. [Joseph A.] Seiss, of Philadelphia, has endeavored to popularize her work on the other side of the Atlantic; and brief references have been made to the subject in such books as *Moses and Geology*, by Dr. Kinns, and in *Primeval Man*; but it was felt, for many reasons, that it was desirable to make another effort to set forth, in a more complete form, the *witness of the stars to prophetic truth*, so necessary in these last days.
>
> To the late Miss Rolleston, however, belongs the honor of collecting a mass of information bearing on this subject; but, published as it was, chiefly in the form of notes, unarranged and unindexed, it was suited only for, but was most

valuable to, the student. It was she who performed the drudgery of collecting the facts presented by Albumazer, the Arab astronomer to the Caliphs of Grenada, 850 A.D.; and the Tables drawn up by Ulugh Beigh, the Tartar prince and astronomer, about 1450 A.D., who gives the Arabian astronomy as it had come down from the earliest times.[1]

Seiss expressed himself similarly:

A more valuable aid to the study of the subject as treated in this volume is Frances Rolleston's *Mazzaroth; or, The Constellations*—a book from an authoress of great linguistic and general literary attainments, whom Providence rarely favored for the collection of important facts and materials, particularly as respects the ancient stellar nomenclature....And from her tables and references the writer of these Lectures was helped to some of his best information, without which this book could hardly have become what it is.[2]

These two later books by Seiss and Bullinger, but particularly Bullinger's, greatly influenced later writers on the subject. Just a few examples of more recent books on the gospel in the stars are Ken Fleming's *God's Voices in the Stars* (1981), William D. Banks' *The Heavens Declare* (1985), and D. James Kennedy's *The Real Meaning of the Zodiac* (1989). There are many others.

The earliest writing on the gospel in the stars is Rolleston's book, so it is the primary source on the subject. The books of Seiss and Bullinger are secondary sources. All other sources generally rely upon Bullinger or Seiss, so they are tertiary sources. Still other sources rely upon even later sources, so they are quaternary sources.

For those who wish to examine the earlier sources for themselves, Seiss' book is most readable. Rolleston's book is the least readable. Part of the problem with Rolleston's book is the very different writing style of a century and a half ago, but also her book was a work in progress. Rolleston died before completion of her manuscript, and much of her manuscript amounted to her notes. Bullinger organized the material a bit better in his book, but it follows much of the style of Rolleston. Seiss' book is organized differently, and it is in a much more readable style. Very few, if any, discussions of the gospel in the

---

[1] E. W. Bullinger, *The Witness of the Stars* (1893; repr., Grand Rapids, MI: Kregel Publications, 1967), iii–iv.

[2] Joseph A. Seiss, *The Gospel in the Stars* (1882; repr., Grand Rapids, MI: Kregel Publications, 1972), 6.

stars are properly referenced, so it is difficult to investigate primary sources on which the gospel in the stars supposedly is based.

Conventional History of the Constellations

There are many constellation systems from around the world, but the system of constellations in the West dominates. Our constellations come from the ancient Greeks. We know of those constellations from Ptolemy, an Alexandrian Greek. Around the year AD 140, Ptolemy wrote *Syntaxis Mathematica*, a 13-volume work on astronomy and mathematics. The *Syntaxis* (as it is more commonly referred to) not only included his understanding of the universe, but also included his understanding of what everyone before him thought about astronomy. We have very few other sources for this information, so if Ptolemy got the history of ancient astronomy wrong, then we have it wrong too.

In the aftermath of the breakup of the Roman Empire, Ptolemy's *Syntaxis* disappeared in the West. Fortunately, after the Arab conquest, Arab astronomers used Ptolemy's work and so they translated it into Arabic. The *Syntaxis* so impressed the Arabs, they referred to it as "the greatest." When the work finally made it back into Western Europe and was translated from Arabic into Latin, it came to be known not by its original title, but by the Latin transliteration of the Arabic for "the greatest," the *Almagest*. Ptolemy's work is historically significant, so rare book collectors frequently have interest in it, but copies that generally sell are late medieval or early renaissance. Note that many are Latin translations of the Arabic translation of the Greek, and that they date more than a millennium after Ptolemy. Earlier copies apparently do not exist. Among the Arabic and Latin manuscripts of the *Almagest* that exist, there are many textual variants, which is common for such manuscripts.

The Almagest discusses 48 constellations, but there is evidence that the constellations are far older than Ptolemy. According to Greek tradition, it was Eudoxus (410/08–357/5 BC) who introduced the constellations to the Greeks from the Egyptians. His work on the constellations was the *Phaenomena*. Although other authors mention Eudoxus and his works, none of the works of Eudoxus have survived. The Greek poet Aratus (315/10–240 BC) wrote a poem by the same name that is loosely based upon Eudoxus' *Phaenomena*. The only surviving work of Hipparchus (190–120 BC) is his critical commentary on the *Phaenomenae* of both Eudoxus and Aratus. The only fragments we have of the text of Eudoxus' *Phaenomena* consist of the quotations in this lone surviving work of Hipparchus.

Aratus' *Phaenomena* poem proved to be very popular in the Greek and Roman worlds. Even the Apostle Paul quoted from it in Acts 17:28. There were several Latin translations of Aratus' *Phaenomena*, the most famous being that of Cicero (106–43 BC). A number of English translations of Aratus' *Phaenomena* exist. In many respects, Aratus' *Phaenomena* is overrated. First, it is very clear that Aratus had very little, if any, knowledge of astronomy, for there are technical astronomical problems with various portions. Second, we must not forget that this is poetry and thus ought not to be treated as a scientific treatise. In similar manner, one would not seriously take Henry Wadsworth Longfellow's poem, "Paul Revere's Ride," as a historically accurate description. Third, there is not much detail contained in Aratus' *Phaenomena*. Contrary to popular belief, only a few names of stars are actually mentioned in the work. The significance of Aratus' *Phaenomena* is that it indicates that the constellations were well established and thus quite old by the third century BC. However, that does not mean that the human recognition of the constellations and understanding of any message in them date from the beginning of creation, as supporters of the gospel in the stars imply. Eudoxus' *Phaenomena* may have taught us much more, but we don't have access to his work.

The *Almagest* contains a catalog of more than a thousand stars. For each star there are coordinates in the sky, a description of the star's position within the outline of its constellation and the star's magnitude (a measurement of how bright the star is). This information allows us to unambiguously locate stars within their relationship to the constellations.

However, there are two problems with Ptolemy's arrangement. First, there are portions of the sky that contain few bright stars, so those regions are not represented by any of Ptolemy's 48 constellations. In the past five centuries, astronomers have added constellations to fill in those gaps. One such constellation is Coma Berenices. Coma, as it often is called, is referenced in ancient and medieval works. For instance, Ptolemy mentioned it, as did Eratosthenes, who wrote about it more than three centuries before Ptolemy, but neither one considered Coma a separate constellation, but rather regarded it as a part of the constellation Leo. Coma as a separate constellation did not start appearing on star charts until the 16th century. As we shall see, Coma is important to the gospel in the stars.

A second problem is that Ptolemy's 48 constellations are restricted to the parts of the sky that were visible from about 35° N latitude a few thousand years ago. There is a large portion of the sky that is not visible from that location. To see those other portions of the sky well, one must travel to the Southern Hemisphere. When Europeans began exploring and then settling portions of the Southern Hemisphere, astronomers soon followed and began creating new constellations in that part of the sky. Of course, being so recent, most modern constellations do not have any sort of lore. Furthermore, many of these newer constellations seem to display a lack of imagination. Modern constellations include devices, such as a clock, an air pump, a compass, a microscope, and a telescope.

Over the years, different astronomers created competing systems of newer constellations. To avoid further confusion, in 1922 the International Astronomical Union (IAU) established 88 modern constellations with defined boundaries. These boundaries encompass the entire celestial sphere, so there is no room for any more constellations. Of course, not all of the competing modern constellations survived, but there are about 40. Most of Ptolemy's original 48 constellations exist in the modern system. The lone exception is Argo Navis, the ship of Jason and the Argonauts in Greek mythology. The IAU adopted the 1752 proposal of the French astronomer Nicolas Louis de Lacaille to subdivide Argo Navis into Carina (the keel, or hull of the ship), Puppis (the deck), and Vela (the sails).

Rolleston was not the first to object to the paganism of the constellations—there had been previous attempts to redefine the constellations in a Christian light. The noted early medieval scholar Bede (672/3–735), an influential English monk and the most learned man in the West in the eighth century, attempted to reassign each of the 12 signs of the zodiac to the 12 apostles. The German lawyer Julius Schiller (1580–1627) attempted much the same thing, but he went much further. In 1627, Schiller published the very beautiful star atlas, *Coelum Stellatum Christianum*. In this Christian star atlas, Schiller not only replaced each of the zodiacal signs with one of the 12 disciples, but he replaced all the constellations then in use with biblical or Christian ones. The new northern hemisphere constellations followed New Testament and early Christian era themes, while the southern hemisphere featured Old Testament themes. Schiller's work never gained any following, so he failed in his attempt to redefine the constellations.

Is There a Biblical Basis for the Gospel in the Stars?

On the surface, the gospel in the stars seems to be plausible, even scholarly, but is this teaching biblically and factually correct? It would seem that such an important "doctrine" (a description used by Seiss) would be clearly taught in Scripture, but nowhere in the Bible are we told the meanings of the various star patterns that we see.

Failing any explicit teaching, at the very least one would expect that elements of the gospel in the stars might have been reinforced in the Bible. For example, the prophet Isaiah in foretelling the Virgin Birth (Isa. 7:14), or Matthew in noting the fulfillment (Matt. 1:23), would have had opportunity to compare to the sign in the sky, but neither did. Or in giving instructions of a proper sacrifice, why did Moses not mention the analogy found in the sky? The New Testament discussions of Jesus' dual nature would have been good opportunities to compare to the celestial counterparts. If the constellations are a God-given revelation, why did the Lord choose not to acknowledge them in His Word?

Claimed Biblical Basis

What is the claimed biblical basis for the gospel in the stars? Proponents of the gospel in the stars theory observe that there was no written revelation for a long time. We do not know when the book of Job was written; some Bible scholars think it likely predates the Pentateuch.[3] Since we do not know when Job was written, let us set it aside for the time being. This would leave the first written biblical text at the time of Moses, in the 15th century BC. If we assume that the Creation was near 4000 BC, there was more than 2,500 years with no written Scripture. Surely, proponents of the gospel in the stars reason, there must have been some mechanism to pass on God's plan of redemption.

Josephus reported that according to Hebrew lore, Adam was the father of astronomical knowledge and that either he or his son Seth created the constellations and passed on that information to their posterity. It is quite likely that Adam developed some astronomical science. After all, Genesis 1:14 records that one of the purposes of the heavenly bodies is for man to mark the passage of time (seasons, days, and years), and this always has been a function of astronomy. As the first man, Adam was in place to establish astronomy, so Josephus may have been right about this. However, Josephus does not tell us in any detail exactly what astronomical knowledge Adam developed. It is reasonable to conclude that Josephus likely had in mind what astronomical knowledge was available in his day.

---

[3] Note that a distinction is made here between the occurrence of the events described in Job, and the composition of the book itself. The events described date to early antiquity, probably to the patriarchal era. However, the composition of the book may well have been later. Job never claims to have been written by an eyewitness.

The proponents of the gospel in the stars hypothesize that God revealed His entire plan of redemption to Adam (or alternately, Seth), and that God ordained the constellations as the mechanism to perpetuate that plan until the giving of the written revelation. However, this is entirely conjectured in that it is not clearly stated or even implied in Josephus and it is not clearly stated in the Bible. Furthermore, this approach seriously underestimates the efficiency of oral transmission of information to reliably preserve truth due to the longevity of the antediluvian patriarchs and the overlapping of generations in the early world.

God Calls the Stars by Name

Proponents of the gospel in the stars further point out that Psalm 147:4 and Isaiah 40:26 tell us that God calls each star by its name. The gospel in the stars proponents then must make two assumptions, though they never clearly state them. The two assumptions are

1. The names that God has assigned the stars (and constellation names) must relate to the primeval gospel.
2. God has shared these names with mankind.

Psalm 147:4 and Isaiah 40:26 make neither of these claims, so these assumptions go far beyond what these verses actually say. Note that neither of these assumptions is supportable by any scriptural text; instead they are conjecture that is necessary for the gospel in the stars to be true.

Romans 10:18

In support of the gospel in the stars, nearly all proponents quote Romans 10:18, which appears to be a direct quote from Psalm 19:4, following the wording of the Septuagint. Romans 10 here is referring to the preaching of the gospel, and verse 18 reads,

> But I ask, have they not heard? Indeed, they have, for

> "Their voice has gone out to all the earth,
>   and their words to the ends of the world."

Proponents of the gospel in the stars reason that since this is a quote from Psalm 19, this necessarily refers to the silent witness of the heavens (stars), and since the gospel message is the context of Romans 10, then this proves that there is a gospel in the stars. However, it does not appear that even a single commentary on the book of Romans supports this understanding of Romans 10:18.

To place this verse into complete context, let us consider verses 12–18:

> For there is no distinction between Jew and Greek; for the same Lord is Lord of all, bestowing his riches on all who call on him. For "everyone who calls on the name of the Lord will be saved."

> How then will they call on him in whom they have not believed? And how are they to believe in him of whom they have never heard? And how are they to hear without someone preaching? And how are they to preach unless they are sent? As it is written, "How beautiful are the feet of those who preach the good news!" But they have not all obeyed the gospel. For Isaiah says, "Lord, who has believed what he has heard from us?" So faith comes from hearing, and hearing through the word of Christ.

> But I ask, have they not heard? Indeed they have, for

> "Their voice has gone out to all the earth,
>   and their words to the ends of the world."

Notice that this passage deals with the gospel presented to both Jews and Gentiles. And the passage raises a series of four rhetorical questions in regard to the gospel. Those questions are:

1. How can people call upon the Lord if they have not believed?
2. How can they believe if they have not heard?
3. How can they hear without a preacher?
4. How can there be preachers if preachers are not sent?

The answers to these rhetorical questions in reverse order are that human preachers must be sent so that people can hear so that they may believe and thus call upon the Lord. Verse 17 claims that faith comes by hearing words—words from the word of God. (An example of the word and preaching working together is the interaction of Philip with the Ethiopian eunuch in Acts 8.) To argue that the very next verse in Romans 10 then refers to a gospel without human preachers, without words, and without the word of God contradicts the intent of the passage. Furthermore, such a teaching places the constellations on an equal footing with Scripture with regard to revelation. To elevate anything to the level of the Bible should make Christians pause.

Granting that Romans 10:18 is a quote of Psalm 19:4, and even if Psalm 19:4 did refer to the gospel in the stars (which we shall soon see is not true), the meaning of any given phrase in one context cannot be used to override the obvious meaning of the same phrase in a completely different context. Furthermore, it is not uncommon for New Testament writers to quote an Old Testament passage and elaborate upon its meaning. In the context of the need for human preachers, the meaning imparted in Romans 10:18 is that even in Paul's time preachers already were spreading the gospel across the known world.

Psalm 19

Let us now examine Psalm 19. There is a very clear division in Psalm 19 between verses 1–6 and 7–14. This division is so stark at first as to almost suggest that this particular psalm may have originally been two psalms that were later joined into one (though there is no evidence of such occurring). However, ancient Hebrew poetry made much use of parallels and contrasts. Placed in juxtaposition, these two passages compare and contrast general revelation and special revelation. The first part of the psalm presents general revelation, opening with the memorable words,

> The heavens declare the glory of God,
>   and the sky above proclaims his handiwork.

The second part refers to special revelation and opens with the equally memorable words,

> The law of the LORD is perfect,
>   reviving the soul.

Both are understood to be revelation—each being a way that God reveals truths to us. That is how they are comparable, but notice how they are contrasted. First, there is a contrast in what the revelation is revealed through. In verses 1–6 the revelation is delivered through the heavens; in verses 7–14 the revelation is delivered through the law, testimony, statutes, commandments, and judgments of the Lord—all synonyms for the Scriptures. General revelation is revealed through the Creation; special revelation is revealed through the Bible.

A second contrast is what kind of truth is conveyed and what that truth does. The heavens show "knowledge" and the "glory of God." General revelation provides broad information about God. In contrast, verses 7 and following indicate that God's word does such things as convert the soul, make one wise, rejoice the heart, and warn one of danger. General revelation might awe us with knowledge of God, but special revelation can transform us with the very nature of God.

The third contrast is how the knowledge is transferred. In the New King James Version Psalm 19:3 reads,

> *There is* no speech nor language
> *Where* their voice is not heard.

There are three words that are in italics, indicating that they are not in the original Hebrew. Italicized words are inserted into the King James and New King James Versions so that it reads better in English. The choice of where words are to be inserted and which words are inserted are editorial decisions made by the translators. There is no debate as to whether the first two italicized words ought to be there in English, for the meaning is not altered if they are there or not, but the first phrase would lack a verb and would read awkwardly otherwise. However, many translators and commentators doubt whether the third italicized word, "where," ought to be there. Omitting "where" (as in the New American Standard Bible) gives a very different read,

> There is no speech, nor are there words;
>   Their voice is not heard.

In other words, the testimony of the heavens is a silent, non-verbal, witness.

In contrast, the Bible gives us the very words of God, dynamically transforming words. Non-verbal communication can convey information, but it lacks precision and specificity, and thus it is very easily misunderstood. The precision of what special revelation can do, as found in verses 7–14, is in stark contrast to the imprecision of what general revelation can accomplish as found in the first six verses. Even in human interaction we frequently communicate by non-verbal means, for body language and facial expressions can convey thoughts. Unfortunately, those non-verbal communications can be tricky to interpret. We easily can misinterpret these silent messages to mean something other than what was intended. A direct verbal statement clearly is preferable to a non-verbal message, as marriage counselors commonly advise when they are trying to help a couple whose marriage is in trouble.

This makes it likely that even if Romans 1:18 alludes to Psalm 19, it is doing so to emphasize this very point. Souls are converted by special revelation, not general revelation. The only way that people can be saved is if God's word is everywhere, and the only way that will happen is if humans spread it everywhere. While general revelation is everywhere, it is silent. Even though it is available in general (to all), it only gives us general information *about* God, not the specific information needed to *know* God. The gospel is not in the stars.

The dichotomy between general and special revelation is also implied in the other passage that concerns general revelation, Romans 1:19–20. Those verses read,

> For what can be known about God is plain to them, because God has shown it to them. For his invisible attributes, namely, his eternal power and divine nature, have been clearly perceived, ever since the creation of the world, in the things that have been made. So they are without excuse.

Notice that Romans 1:20 states that there are two things that general revelation tells us, "his eternal power and Godhead." That is, God exists and is very powerful. Romans 1 also tells us that men are without excuse for their condition, but there is nothing in general revelation that tells us that God sent His Son into the world to pay that penalty for our sin. To learn these and other things related to salvation, we must turn to special revelation, the Bible. In other words, general revelation can lead us to conclude that there is a Creator and what at least some of His attributes are, but general revelation alone is insufficient to lead us to Christ. Furthermore, this proscription from Romans 1:20 would seem to rule out the entire gospel message being found in the stars and constellations (general revelation) as supporters of the gospel in the stars require.

Genesis 15:5

Some supporters of the gospel in the stars claim that Genesis 15:5, when properly interpreted in light of Galatians 3:16, teaches the gospel in the stars. Galatians 3:16 reads (in the New King James Version),

> Now to Abraham and his Seed were the promises made. He does not say, "And to seeds," as of many, but as of one, "And to your Seed," who is Christ.

They argue that this is a direct reference to God's promise to Abraham in Genesis 15:5, which reads (in the King James Version),

> And he brought him forth abroad, and said, Look now toward heaven, and tell the stars, if thou be able to number them: and he said unto him, So shall thy seed be.

The Hebrew word translated "seed" above in Genesis 15:5 is a collective noun, so depending on the context, Bible scholars will translate it as either a singular or plural (in the latter case often with the word *descendants*). Supporters of the gospel in the stars reason that since Galatians 3:16 explicitly ties this to the singular in Jesus Christ, then the seed of Genesis 15:5 ought to be interpreted in terms of the singular in Jesus as well.

Supporters of the gospel in the stars also note that the Hebrew word סָפַר (*sāp̄ar*) appears twice in Genesis 15:5 translated in the King James Version first as "tell" and then as "number." They further note that this Hebrew word can have two different meanings, either "to count" numerically or "to tell," as in telling a story. Indeed, *sāp̄ar* is translated into English as "declare," "speak," and similar words a number of times in the Old Testament. But *sāp̄ar* also is translated as "number" or "count" many times in the Old Testament. As with any passage, the context is important in translating this properly. While today the King James Version appears to make a distinction in the two uses of *sāp̄ar* in Genesis 15:5, to early 17th-century readers it probably did not. Indeed, the English Standard Version translates the word as "number" both times, the New King James Version and the New International Version render the word "count" in both instances, and the New American Standard translates it first as "count" and then "number."

How do some supporters of the gospel in the stars interpret Genesis 15:5? The verse refers to the miraculous promise to an old man with an old, barren wife and without an heir that God would make his descendants so numerous as to be uncountable; but supporters of the gospel in the stars find a different meaning. They claim that God told Abraham to look at the stars and retell the story found in the stars. As Abraham recounted the story of redemption found in the stars that had been handed down to him, God informed Abraham that this was to be the story of Abraham's seed.

There are at least two things wrong with this interpretation. First, no commentators on either Genesis or Galatians endorse this interpretation of the supporters of the gospel in the stars. Second, Galatians 3:16 is not an exclusive reference to Genesis 15:5. For instance, it appears to be a better fit to Genesis 12:7, the first promise concerning Abraham's seed. Genesis 12:7 reads,

> Then the LORD appeared to Abram and said, "To your offspring [seed] I will give this land." So he built there an altar to the LORD, who had appeared to him.

Compare the wording of Genesis 12:7, 15:5, and Galatians 3:16, and you will see that Galatians 3:16 more closely reflects Genesis 12:7, not Genesis 15:5.

The word *sāpar* does not appear in the Hebrew of Genesis 12:7, nor are the stars mentioned. For that matter, the innumerable nature of Abraham's seed is not mentioned here either. Since the phrasing of Genesis 15:5 is different, it is strained to insist that Galatians 3:16 must refer exclusively to Genesis 15:5. Of course, the promise of Genesis 12:7 is repeated and expanded in Genesis 13:14–16 and again in Genesis 15:5, both of which mention the innumerable nature of Abraham's seed. These themes also are repeated in Genesis 22:17 and Genesis 26:4. Genesis 22:17 promised that Abraham's seed would bless all nations. Galatians 3:14 speaks of the blessing through Abraham extended to the Gentiles, which echoes Genesis 22:17 (cf. 12:3) and frames the context of Galatians 3:16. By concentrating on only Genesis 15:5 and excluding the other four relevant verses mentioned here, supporters of the gospel in the stars misinterpret Genesis 15:5.

Genesis 1:14

Another passage used to support the gospel in the stars is Genesis 1:14, which tells us that one of the purposes for the celestial lights is to be for signs. What does it mean for astronomical bodies to be for signs? Supporters of the gospel in the stars theory generally believe that this must refer to the gospel message. However, there are several *biblical* answers for what these signs may be.

First, in Matthew 16:1–4, the Pharisees asked Jesus for a sign. He responded by quoting from some of their own teachings about the sign of a red appearance in the sky to forecast weather, but chided them for not recognizing the signs of the times. Thus, in context, the people well versed in the Old Testament understood that this form of weather forecasting was a kind of sign. Second, as Psalm 8, Psalm 19, and Romans 1:18–20 tell us, God's existence is revealed through the heavens, constituting a sort of sign. Third, the star that led the magi to the infant Jesus (Matt. 2:1–2, 9–10) was undoubtedly a sign from heaven. Fourth, there will be signs in heaven that reveal God's wrath (Isa. 13:9–13; Joel 2:30–31; Matt. 24:29–31; Mark 13:24–27; Luke 21:25–28; Rev. 6:12–17).

Thus, in other biblical passages we have at least four types of possible signs concerning the heavens and the heavenly bodies that conform to the God-ordained purpose for them. With no clear biblical support for these signs being the gospel in the stars, it is pure conjecture that Genesis 1:14 requires that there be a gospel in the stars.

Conclusion

In view of what precedes, we must ask, How is the gospel in the stars different from the dual-revelation theory, the idea that God has revealed Himself in the natural world in the same way (and with the same clarity) that He has revealed Himself in Scripture? Some erroneously elevate the role of the witness of creation, claiming that it amounts to a 67th book of the Bible that reveals more than it actually does about God's attributes and purposes. How is the gospel in the stars different from this alleged but untrue idea of the 67th book?

There is also the problem of certain things being made to have significance before their time. Despite Isaiah's foretelling, the Virgin Birth was not fully understood until the writing of the Gospels. In a similar way, despite Jesus' prediction of His death and Resurrection, His followers did not understand until after it had happened. Before the Crucifixion and Resurrection, what was the significance of the Cross? None, but supporters of the gospel in the stars attach much predictive significance to Crux, the Southern Cross. So what role did it play? It could not have been a part of an Old Testament gospel—crucifixion was a much later invention widely used by the Romans, and so would have meant nothing to the patriarchs. And what kind of prediction would it have been, given that no New Testament allusion is made to it? The Gospels, particularly Matthew's, clearly note fulfillment of Old Testament predictions, but a cross in the sky is never mentioned.

In summation, the supposed biblical support for the gospel in the stars is very weak, amounting to a very oblique argument at best. And some texts used in support are taken out of context or misapplied. Supporters of the gospel in the stars argue certain biblical passages support their view, but no commentaries on those passages agree with the interpretations offered. These novel interpretations date no earlier than Rolleston's work in 1865. When we find some understanding of a biblical passage that no one before has seen, it likely is because that meaning is not there.

CHAPTER 15

The Gospel in the Stars:
Assumptions and Some Difficulties

Most treatments of the gospel in the stars suppose that Francis Rolleston found clear evidence of the gospel in the stars by reading certain ancient texts. However, this is not the case. Rather, motivated by her desire to save star lore from its pagan roots, Rolleston developed her conjecture regarding the origin of the constellations. In her book, Rolleston discussed the assumptions that she used to interpret star and constellation names.[1]

What Are the Assumptions?

Let us summarize the assumptions that supporters of the gospel in the stars theory must make.

1. God not only named the stars (Ps. 147:4; Isa. 40:26), but He shared those names with man (Adam).
2. God's names for the stars convey the gospel message.
3. There was a need for the gospel message prior to the giving of that message in the written word of God.
4. The original language of man was Hebrew.
5. At the time of Babel, only pronunciation changed; thus Hebrew roots were preserved in all languages.
6. Star names that we have today are ancient in origin, dating from the earliest times and thus reflect the early meanings delivered by God.

---

[1] In this chapter, I will discuss various constellations. To view star charts of the constellations, please see Danny R. Faulkner's *The New Astronomy Book* (Green Forest, AR: Master Books, 2014), 90–91.

There is an alternative to point 2, that God revealed the gospel message to Adam and that either he and/or Seth named the stars and constellations to reflect that message. This is less popular among supporters of the gospel in the stars theory, but they frequently quote Josephus concerning Adam and the origin of the constellations. Josephus gave credit to Adam and/or Seth apart from any mention of God's help.

Note that all six assumptions must be true for the gospel in the stars theory to be true. None of these assumptions can be proven, so they truly are assumptions. If any of the six assumptions is not true, then the gospel in the stars is not true. What about these assumptions?

The first assumption asserts that God has shared the names for the stars with us, but as we saw earlier, there is no scriptural basis for this belief. The two verses cited (Ps. 147:4; Isa. 40:22) merely state that God has named the stars (and the verses imply all the stars), and we know this through the specific information given in these two divinely inspired verses. But to conclude that these two verses necessarily imply that God shared those names with Adam goes far beyond what the verses say and reads too much into them. Since no names of individual stars are in the Bible, the idea that God shared any of His names for the stars with man must necessarily be extrabiblical. If such extrabiblical revelation existed, all sorts of questions about the nature of special revelation and the preservation of that revelation arise.

For instance, one might question what other extrabiblical writings or traditions were inspired and why they have not been preserved. These questions can erode confidence in the doctrine of revelation. Interestingly, while no individual stars are unambiguously named in the Bible, there are names of a few *groups* of stars mentioned in the Bible (for example, Orion and the Pleiades), but advocates of the gospel in the stars tend to ignore those, opting instead to find meaning in Hebrew words for the non-biblical (and non-Hebrew) names. But this is inconsistent, for if God has ordained certain names for groups of stars, does it not stand to reason that He would use those names in His divine revelation, the Bible? Then why not search for meaning in those biblical names rather than search for meanings in non-biblical names?

The second assumption is the key one concerning the gospel in the stars—that God's names for the stars convey the gospel message. No biblical passage clearly teaches this. We saw in Chapter 14 that the attempted proof texting for this idea reads far too much into various passages and has no support from commentators.

The third assumption is related to the very old question concerning the fate of those who have never heard the gospel as explained in the New Testament. Though this has been discussed in numerous places, there is no totally satisfactory answer to this question. It is important to emphasize that

1. Every person who has ever lived has had the witness of creation and the witness of conscience (Job 12:7–10; Ps. 19:1 [cf. Rom. 10:18]; 97:6; Acts 14:15–17; 17:24–29; Rom. 1:18–20; 2:14–15).
2. If a person responds positively to the truth he has, God will get more truth to him, even miraculously if necessary (e.g., Acts 10:1–5).
3. No one deserves to have more truth than creation and conscience supplies, and any more truth that anyone does receive is the result of grace. God is just, even if He gives no more truth as Romans 1:20 makes clear.

As for the transmission of a salvation message prior to the written word of God, how did any of God's instruction to man come down to the patriarchs? We know that God directly spoke to certain individuals such as Adam, Cain, Noah, Abraham, and Moses. God may have directly revealed Himself to any number of other individuals not recorded in Scripture. Furthermore, we cannot discount the oral and even written testimony (though not inspired as with Scripture) of followers of God. These are only a few possible ways that this information from God could have been conveyed apart from a gospel in the stars. And keep in mind that the gospel in the stars explanation for this question dates no earlier than 1865.

The fourth assumption about Hebrew or some form of Hebrew being the primordial language has the greatest indirect support of these six assumptions, and it enjoys a broad range of support among Christians. However, this is hotly disputed by many with a background in Hebrew or in linguistics, as it is well known that biblical Hebrew is a language that underwent extensive development. The indirect argument for this is two-fold. First, the names of many of the patriarchs convey information in Hebrew. It seems reasonable that these names must have had meaning in the original language. However, there is no reason that those names themselves were not translated into Hebrew from the original language.

The second indirect support is the so-called tablet model of Genesis, a theory popular with a number of creationists.[2] This view is based on the 11

---

[2] However, there is controversy surrounding this theory, nor is it widely accepted among conservative Christians.

occurrences of the Hebrew word תּוֹלְדוֹת (*tôlēdôt*, translated "these are the generations" or "this is the account") scattered through Genesis. It is proposed that this indicates that Moses wrote Genesis on the basis of accurate written documents (perhaps written by the individuals whose names are associated with each *tôlēdôt*) passed down through the patriarchs from Adam to Moses. Key patriarchs kept a history of their lineage, and those same patriarchs added their own stories to that history.

If the pre-Babel writings of the patriarchs were in some language unknown to Moses, then he could not have collated those records into the book of Genesis. On the other hand, some Christians believe that Sumerian may have been the pre-Babel language. The Sumerian civilization is the earliest known large civilization, and the Sumerian language is a language isolate. That is, it is a language with no known relatives. This suggests that it might have been the pre-Babel dispersement language. On the other hand, the confusion of languages at Babel may have resulted in no one speaking the original Adamic language.

The fifth assumption, that only pronunciation was changed at Babel, is explicitly stated by Rolleston. In the explanations of her tables, Rolleston stated,

> The names are here explained on the supposition that the first language was given by the Creator to the first man, conveying ideas to the mind by sounds, as impressions of form and colour are conveyed by sight. In all languages these sounds are traceable, conveying the same ideas. In the dialects of the most ancient and earliest civilized nations they are the most recognizable: in those the most barbarous the most obscured. This primitive language appears to have been spoken by Noah, from the names given by him to his sons. In the confusion of the lip at Babel, pronunciation, and not words or roots, were altered. This may be inferred from the presence of the Hebrew roots in the dialects of all nations.[3]

Thus, Rolleston assumed that Hebrew is the closest language to that of Adam. As stated above, this is a common belief among creationists, but it is not necessarily true. As for her assumption that only pronunciation changed at Babel, qualified Christian linguists find no support for this assumption.

The sixth assumption, that the star names we have date from the beginning of creation, is very doubtful. The earliest documentation of only a few star

---

[3] Frances Rolleston, Mazzaroth (1875; repr., York Beach, ME: Weiser Books, 2001), part 2, p. 3.

names goes back to Aratus in the third century BC. Most star names have documentation that is medieval in origin. We can document that some star names are of very recent origin.

What Is the Methodology?

With these six assumptions, Rolleston used this methodology to interpret the names of stars:

1. The original meanings of star names are determined by identifying homophones and meanings in Hebrew and other Semitic languages.
2. Vowels can be ignored, because of differences in pronunciation and because vowel points were added to Hebrew much later.

Rolleston searched for homophones in Hebrew to match star and constellation names.[4] For instance, Rolleston reasoned that Latin derived from Etruscan, and she in turn thought that the Etruscan people came from Assyria. Since Assyrians spoke Semitic languages, Rolleston concluded that the Etruscan language was Semitic. Thus, Rolleston thought that she could find meanings of Latin names from Hebrew roots. However, it is doubtful that the Etruscans came from Assyria and hence that their language was Semitic. In some cases, Rolleston claimed to find root meanings in other Semitic languages. One could only guess that she resorted to this when she found no satisfactory match to any Hebrew word.

But why stop with star names? It would seem that one could trace the meanings of nearly all words in all languages in this manner. Given the highly speculative nature of this approach, her conclusions on particular meanings from Hebrew and related Semitic languages are suspect at best. Yet, there is no doubt expressed in her assertions about root meanings, nor in the claims of those who have followed her lead. I will now illustrate the problems with Rolleston's approach by examining some of the claims made about more technical issues. Chapter 16 will discuss broader issues and other difficulties with the gospel in the stars.

The Constellations According to the Gospel in the Stars

As previously mentioned, most of what we know of ancient astronomy comes from the Almagest, a second-century AD book written by Claudius Ptolemy. Ptolemy described 48 constellations, 12 along the zodiac with 36 others.

---

[4] Ibid, part 2, pp. 1–2.

Supporters of the gospel in the stars preserve this number, but they do so while deleting three constellations and replacing them with three other constellations that allegedly predate Ptolemy. The reason for this is not clear. The deleted constellations are Corona Australis (the Southern Crown), Equuleus (the Small Horse), and Triangulum (the Triangle). The replacements are the Bands (binding together Pisces the Fish), Coma Berenices (Bernice's Hair), and Crux (the Southern Cross). The latter two are now recognized as constellations, but are of more recent origin. Rolleston claimed to have found an ancient source that separated the fish and the bindings into two separate constellations, so apparently she decided that this was a primordial constellation.

Advocates of the gospel in the stars arrange their 48 constellations into decans. According to the gospel in the stars theory, a decan is three non-zodiacal constellations associated with a zodiacal sign. That is, each of the 12 signs of the zodiac has three associated constellations. Hence, all 48 constellations are neatly contained within the 12 decans. Before discussing Crux and Coma Berenices in more detail, let us first examine the decans, as defined by advocates of the gospel in the stars.

Decans

Although many ancient cultures referred to decans, they defined decans very differently from the way Rolleston used the term. These cultures divided each astronomical sign into three decans. There are 12 signs circling the sky, with each sign stretching over approximately 30° of the ecliptic. Thus, each decan spans roughly 10° along the ecliptic. Since it takes approximately 360 days for the sun to complete a circuit with respect to the stars, the sun occupies each decan for roughly ten days. In fact, the word decan derives from the Latin and Greek roots from which we get the word *decade,* meaning "ten."

However, supporters of the gospel in the stars reject this logical and straightforward etymology of decan. Rolleston wrote that the word comes from the Hebrew and Arabic word דקק, meaning "to break into pieces."[5] However, this word actually is Aramaic. Seiss went on to claim that the word *deck,* as on a ship, comes from the same root.[6] This is not correct, as any good dictionary traces the word *deck* back through Dutch to German to Latin and Greek from a word meaning "to cover."

---

[5] Ibid, part 2, p. 14.

[6] Joseph A. Seiss, *The Gospel in the Stars* (1882; repr., Grand Rapids, MI: Kregel Publications, 1972), 18.

In most ancient cultures each decan was ruled by some other astronomical body. In some systems, as in ancient Egypt, each decan was ruled by a particular star that rose with the decan. In other systems, it was the sun, moon, or one of the five naked-eye planets. Decans are directly connected to ancient astrology; in modern times, the decans have fallen into disuse among astrologers. One of the more recent discussions of decans is that of William Lilly in his three-volume work *Christian Astrology*, originally published in 1647. In place of the more modern term decan, Lilly used the terms *decanate, decurie,* or *face.*

Where did this understanding of decans unique to the gospel in the stars come from? It appears to have originated with Rolleston and not with any ancient astronomers. Rolleston stated, "The Decans are here arranged from a work by Albumazer, Flor. Astro., a Latin translation of which is in the Library of the British Museum."[7] Later, Rolleston mentions this work again and there offered quotes from Albumasar (the modern spelling of the name).[8] This portion of her book is a bit disorganized (she died before this portion was completed, according to an "advertisement" inserted at the beginning of part IV), so it is not entirely clear whether the quotes offered here are indeed from this particular work of Albumasar or some other (Rolleston mentioned no others). Rolleston probably did her own translation of this work from the British Museum Library copy.

The "Flor. Astro." must be the *Florum Astrologie,* or, in English, "The Flowers of Astrology," though this work is better known as *Liber Florum,* or *Book of Flowers,* a treatise on mundane astrology. The "flowers" in the title refers to "choice selections" rather than to plants. Fortunately, there is a 2008 English translation of the Latin translation of *The Book of Flowers* by James H. Holden. This translation has no description of decans in it, let alone the arrangement that Rolleston presented. For that matter, no non-zodiacal constellations are even mentioned. Therefore, it is a mystery as to where Rolleston's arrangement of these decans came from. Given the generally poor manner that Rolleston handled sources and her ability to create false history, it is likely that Rolleston may have misunderstood a portion of Albumasar and that she essentially created the arrangement of the decans herself.

How might Rolleston have created these decans? There is a possible scenario. *The Book of Flowers* mentions triplicities of four of the zodiacal constellations.

---

[7] Frances Rolleston, *Mazzaroth,* part 2, p. 14.
[8] Ibid, part 4, p. 12.

For instance, it states, "When Saturn is lord of the year and in Taurus or its triplicity…" Perhaps Rolleston thought that the triplicities referred to three ancillary constellations that were applied to each of the zodiacal constellations. If this is how Rolleston arrived at her decan designations, then she completely misunderstood what Albumasar was saying. Albumasar mentioned only the first four zodiacal constellations, Aries, Taurus, Gemini, and Cancer, in the context of triplicities. This is because each of these four signs had two other zodiacal signs assigned to its triplicity. Astrologers divide the 12 signs into four groups corresponding to the four ancient elements: fire, earth, air, and water. The fiery signs are Aries, Leo, and Sagittarius; the earth signs are Taurus, Virgo, and Capricornus; the air signs are Gemini, Libra, and Aquarius; the water signs are Cancer, Scorpius, and Pisces. Each of these four groups consists of three triplicities; this is what is meant by a triplicity.

Yet another problem is how Rolleston knew which three constellations were to be combined with each zodiacal sign. For instance, how did she know that the two bears and Argo Navis (the one now defunct Ptolemaic constellation) were to be associated with Cancer? Rolleston stated that "the three decans attributed to each sign come to the meridian with it."[9] Therefore, Rolleston likely determined when some prominent portion of each of the other 36 constellations crossed the meridian along with each zodiacal constellation during some ancient epoch.

Seiss rearranged the quotes allegedly from Albumasar and added an additional quote from another source. Seiss normally referenced quotes, but he did not reference the Albumasar quotes, so it is not clear if he checked these supposed quotes himself or merely relied upon Rolleston to correctly quote Albumasar. If one uncritically reads what Seiss wrote here about decans, it is convincing. However, once one realizes that there is no basis for the decanal arrangements as put forth by gospel in the stars advocates, then the quotes do not amount to much. That is, one could easily understand these quotes in the context of the proper view of the decans being $10°$ increments within each zodiacal sign. In fact, Seiss concludes his discussion of the decans with this interesting sentence:

> And after the closest scrutiny, those who have most thoroughly examined and mastered the subject in its various relations entirely agree with the same enumeration, which I therefore accept and adopt for the present inquiries into this starry lore, sure that the particular examination of each sign, with

---

[9] Ibid, part 5, p. 15.

the Decans thus assigned to it, will furnish ample internal proof that this enumeration is correct according to the original intention.[10]

Here Seiss appeals to self-consistency for ultimate proof of the arrangement. That is, the three other constellations supposedly associated with each zodiacal sign complement each other so well as to demonstrate that the arrangement is true. Given how similar to a Rorschach test that this appears to be, one probably could find connections in any number of possible combinations of constellations. However, Seiss' statement here appears to be a tacit admission of how poorly founded the arrangement of decans set forth in his book actually is.

Note that this scenario explaining how Rolleston might have established her system of decans is conjecture. There is no precedent for Rolleston's decans in the literature, and this arrangement appears to be unique to the gospel in the stars, suggesting that this arrangement originated with her. It is regrettable that all gospel in the stars advocates have uncritically accepted Rolleston's decans as established ancient practice.

Crux (the Southern Cross)

In ancient times, the stars of Crux were included within the constellation Centaurus. For instance, Ptolemy included the stars of Crux in his catalog, but as part of Centaurus. Crux as a separate constellation is sometimes attributed to Augustin Royer in 1679, but Jakob Bartsch listed it separately in 1624, and Emerie Mollineux illustrated Crux on a star globe as early as 1592. On the other hand, Johann Bayer followed the ancient custom of including the stars of Crux within Centaurus in his 1603 work *Uranometria*.

The attempted identification of symbols of the Cross throughout Christian history in support of some primeval gospel is not unique or even new to Rolleston, as evidenced by William Haslam's 1849 book, *The Cross and the Serpent*. This book or ones similar to it likely influenced Rolleston. Since the Cross is a symbol long used by many Christians, Crux could have an obvious connection to a salvation message. Therefore, the addition of Crux fits with the gospel in the stars message, if Crux were an ancient constellation. Supporters of the gospel in the stars point out that Crux was low in the sky at temperate latitudes in ancient times, but an effect called precession now has rendered it visible only at much more southern latitudes. Supposedly more ancient catalogs

---

[10] Joseph A. Seiss, *The Gospel in the Stars*, 18.

included Crux, but in his lifetime Ptolemy could not see it, so Ptolemy deleted it from his catalog. It is further asserted that there was a faint memory of Crux, and as navigators sailed southward five centuries ago they rediscovered it.

What of the claim that Ptolemy could not see Crux? It is true that this part of the sky was once visible from the latitudes where the ancients observed, but precession has caused this part of the sky to be lost at these latitudes today. However, the claim that the stars of the Southern Cross would not have been visible to Ptolemy is false. When Ptolemy observed in Alexandria about AD 140, Alpha Crucis, the southernmost star of the Southern Cross, reached about 9° above the horizon. The other stars of the Southern Cross were even higher, making this group of stars quite visible with little difficulty, so they hardly were a faint memory at the time of Ptolemy. Hence, not only is there no evidence for this false history of Crux, the facts clearly argue against it, for Ptolemy certainly did see the stars that make up Crux, and he included them as part of the constellation Centaurus in his catalog. Apparently Rolleston began this false history, and those who have followed her have uncritically accepted what she wrote.

This criticism (among a few others) apparently was lodged against the gospel in the stars thesis shortly after Rolleston proposed it, prompting Seiss to insert a supplement in his book where he dealt with this and other questions. Seiss appealed to Ptolemy to justify his argument about Crux:

> Ptolemy himself also confesses that in the tables and charts presented by him liberties were taken to change figures and the places of stars in them.... Whether, therefore, the Southern Cross belongs to the ancient forty-eight constellations or not cannot be determined from its absence from the Ptolemaic tables, as that can argue nothing for or against the assertion that it does so belong, apart from other showings.[11]

This is a remarkable admission. No one can prove that Ptolemy deleted the Southern Cross, but Ptolemy might have. There is no evidence brought to support this false history; rather, it was simply conjecture, using Ptolemy's vague statement as justification.

Coma Berenices

Most of the stars of Coma Berenices are in the star cluster Melotte 111. Because it is so close to us, Melotte 111 extends more than 7° across the sky. The stars are faint, so on a dark, clear night the cluster's stars appear hazy, suggesting

---

[11] Ibid, 170.

a hairy appearance. The Coma Berenices constellation is the hair of Queen Berenice II of Alexandria. Since she died in 221 BC, that constellation cannot date any earlier than that. References to this constellation being the hair of Queen Berenice began appearing within a century after her death. Ptolemy mentioned this faint grouping of stars as hair, but did not ascribe it to her. Rather, Ptolemy included the stars of Coma with Leo, as the tuft on the end of the tail of the lion. The stars of Coma were not removed from Leo and made a separate constellation until much later. Uranometria included Coma as a separate constellation, but it was proposed as a separate constellation by Tycho Brahe a few years earlier than Bayer.

Seiss and Bullinger identified Coma Berenices as a woman holding a small child (an obvious reference to Mary and Jesus), even including identical drawings of this grouping. The source of the drawing appears to be the Dendera planisphere, a stone star chart found in Dendera, Egypt, which is about 2,000 years old (though at the time of Rolleston, Seiss, and Bullinger, it was thought to be far older). Supporters of the gospel in the stars claim that the original name of the constellation was Coma, but that later cultures misunderstood this. For instance, Seiss opined, "The Greeks knew not how to translate it, and hence took *Coma* in the sense of their own language, and called it *hair— Berenice's Hair*."[12]

Bullinger had similar reasoning.[13] The methodology of Rolleston was to look for Hebrew words in various names, and the intended Hebrew word root is כמה (*kmh*), supposedly meaning "the desired, or longed for." However, *kmh* is a verb, not a noun. Furthermore, it appears that both Bullinger and Seiss relied upon Rolleston for this, but misunderstood what Rolleston actually said. Rolleston noted that on the Dendera planisphere there is a figure of a woman holding a small child below the figure of Virgo, and she surmised that this otherwise separate drawing was related to Virgo, though Virgo has no child displayed with her.[14] Rolleston thought that Coma represented the branch or sheaf of grain that Virgo normally is depicted as holding.[15] One could get that understanding from the Dendera planisphere, for the scale is difficult to interpret, and the fuzzy appearance of the Coma star cluster could be said to resemble a sheaf of grain.

---

[12] Ibid, 29.

[13] E. W. Bullinger, *The Witness of the Stars* (1893; repr., Grand Rapids, MI: Kregel Publications, 1967), 35.

[14] Frances Rolleston, *Mazzaroth*, part 2, p. 16.

[15] Ibid, part 2, p. 17.

Incidentally, Rolleston consistently refers to the sheaf as a branch in an obvious connection to Isaiah 11:1. However, that branch is from a stump of a tree, and Virgo always is depicted with a sheaf of grain, not a tree branch.

Are we to equate the modern constellation of Coma Berenices with the depiction of a woman holding a child on the Dendera planisphere? Hardly. The figure in question on the Dendera planisphere is below both Virgo and Leo, but Coma Berenices is above Virgo. The star charts of Seiss and Bullinger (clearly show this little constellation of a woman holding a child above Virgo that is claimed to be copied from the Dendera planisphere. Although this is the correct location of Coma Berenices, it is clearly on the other side of Virgo where the Dendera planisphere depicts it.[16] Obviously, whatever the Dendera planisphere is depicting, it is not to be identified with the constellation Coma Berenices as Bullinger and Seiss have done, because it is in the wrong location.

As previously mentioned, Rolleston appeared to get the location of this image correct, so why did Bullinger and Seiss confuse this? Elsewhere on the same page, Rolleston states under the list of the first decan, "COMA, the Branch or Infant near or held by the Woman."[17] This contradicts what Rolleston wrote later on that page, and does seem clearly to imply the equivalence of Coma Berenices with this supposed constellation of the mother and child. So this apparently is the source of the misunderstanding by Seiss and Bullinger. This also apparently is how Rolleston, Seiss, and Bullinger managed to conjecture a supposedly ancient constellation, "The Desired For," out of Coma Berenices, although there is absolutely no evidence that such a constellation existed in ancient times. Given the connection to the Virgin Birth, this particular constellation is very important to many supporters of the gospel in the stars today. It is most unfortunate that they have uncritically accepted Rolleston, Seiss, and Bullinger on its existence.

Rolleston quoted Albumasar as saying,

"There arises in the First Decan, as the Persians, Chaldeans, and the Egyptians, the two Hermes and Ascalius teach, a young woman, whose Persian name translated into Arabic is Adrenedefa, a pure and immaculate virgin, holding in the hand two ears of corn, sitting on a throne, nourishing an infant, in the act of feeding him, who has a Hebrew name (the boy, I say), by some nations named Ihesu, with the signification Ieza, which in Greek call Christ."[18]

[16] Ibid, part 5, p. 1.
[17] Ibid, part 2, p. 16.
[18] Ibid, part 2, p. 17.

Rolleston went on to comment that "Ieza" probably was "the Hebrew verb yesha, [meaning] to save." However, this verb never appears in the *Qal* stem at all, nor would it mean "to save" except in the *Hiphil* stem. The root that Rolleston had in mind is יָשַׁע (*yšʿ*), but it does not appear that Rolleston knew what she was talking about here. She also footnoted that "Adrenedefa" was from Hebrew, meaning "a pure virgin, offering," with Exodus 35:29 as a reference. The intended Hebrew word here is נְדָבָה (*nᵉdābâ*), which means "free will offering," but it is not a good fit. And how this relates to a virgin is unknown, though it possibly may mean that Rolleston thought that Mary made a free will decision to remain a virgin as a form of sacrifice, for Rolleston believed in the perpetual virginity of Mary (contra Matt. 1:25; 13:55–56; etc.).[19] Unfortunately, Rolleston did not give a reference to where in Albumasar she found this quote, so it is not possible to check this quote for accuracy. It is not found in *The Book of Flowers*, the only Albumasar work actually mentioned in Rolleston's book, so there is considerable doubt that it is a legitimate quote. Did Rolleston fabricate this quote? This is such a wonderful quote to support the gospel in the stars thesis that most who followed Rolleston have uncritically repeated the quote, even embellishing it by explicitly pointing out that Albumasar was Muslim, not Christian.

Rolleston had a poor track record in accuracy and documentation, so we ought to be very skeptical of this quote. Assuming for now that it is an accurate quote, does it make the strong case that gospel in the stars advocates think? Hardly. First, contrary to common belief among Christians, Muslims do not doubt the Virgin Birth of Jesus. Muslims view Jesus as among the greatest of the prophets, second only to Mohammed. Mohammed himself taught that Jesus had no earthly father, but that it does not follow that Jesus is Immanuel. So this quote, if legitimate, is not a grudging admittance by a Muslim as many seem to think.

Second, Albumasar wrote in the ninth century, eight centuries after the ministry of Jesus. Rolleston relied upon a Latin translation made at least six centuries after Albumasar. This leaves a tremendous amount of time for Albumasar to have been influenced by Christian teachings and for later transmission of his writings to have been influenced by Christian teachings. Rolleston assumed that Albumasar was transmitting ancient, pre-Christian thought, but this has not been demonstrated. Without clear demonstration of

---

[19] Ibid, part 2, pp. 98–99.

the clear antiquity of what Albumasar allegedly wrote on this matter, Rolleston's argument does not prove anything on this matter.

While on this subject, we ought to address a common misunderstanding found in the literature (including gospel in the stars literature) concerning Virgo. Many sources quote a line from William Shakespeare's *Titus Andronicus*, Act 4, Scene 3 to prove that as recently as Shakespeare's time people generally pictured Virgo with a young child in her lap (an obvious reference to Mary and young Jesus), with the implication that this supposedly ancient depiction has since disappeared. However, many depictions of Virgo that are earlier than or are contemporary to Shakespeare show no such thing. Perhaps we are supposed to believe that this inferred ancient depiction disappeared, and then briefly reappeared four centuries ago, only to disappear once again.

The context of the line in the play is a scene in which the characters are shooting arrows up into the sky, arrows with messages to the gods attached. Titus observes that one of the arrows was shot up to Virgo. Rolleston stated that the arrow was shot up to "the good boy in Virgo's lap."[20] Seiss quoted the line, "to the good boy in Virgo's lap," and Bullinger rendered it, "Good boy in Virgo's lap."[21]

Notice that these quotes do not exactly agree, but this at least can be attributed to various versions of *Titus Andronicus*, for versions differ in punctuation, spelling, and even words. These quotes, taken in isolation, could be interpreted to refer to a boy sitting in Virgo's lap, but this is not what the passage means in context. Here is the entire line by Titus: "Oh, well said, Lucius! Good boy, in Virgo's lap! Give it Pallas."

Here Titus is praising his grandson, Lucius, for his excellent shooting. Earlier, Titus had affectionately called young Lucius "boy" more than once, and "well said" is better understood today as "well done." Titus congratulates Lucius for squarely landing an arrow in Virgo's lap, so the "good boy" here refers to Lucius, not to baby Jesus. Titus goes on to praise his nephew Publius for his arrow shooting off one of the horns of Taurus. Titus' brother, Marcus, and father to Publius, further elaborates on the jesting by noting that when Publius' arrow struck Taurus, the bull knocked Aries so that both of Aries' horns fell to the earth. This may appear as nitpicking to some, but this incorrect interpretation of Shakespeare illustrates how proponents of the gospel in the stars so easily misread texts in support of their thesis.

[20] Ibid, part 2, p. 17.
[21] Joseph A. Seiss, *The Gospel in the Stars*, 28; E. W. Bullinger, *The Witness of the Stars*, 36.

Problems with the Interpretation of Orion

The name "Orion" appears three times in the Bible (Job 9:9; 38:31; Amos 5:8). Rolleston correctly noted that כְּסִיל (k^esîl) is the Hebrew word translated as "Orion" in all three instances.[22] Rolleston viewed Orion as a type of Christ. How does one establish Orion as a type of Christ? Part of this comes from the name "Orion." The meaning and source of this name is obscure, but it likely comes from Akkadian and means "light of heaven"; so proponents of the gospel in the stars have in mind Matthew 4:16 in their connection. Furthermore, although on most charts Orion's foot rests upon a hare, Rolleston claimed that on at least one ancient Indian star chart there is a snake in place of the hare. Presumably, this snake has bitten, or bruised, Orion's heal, but Orion is crushing the serpent's head in fulfillment of the first Messianic prophecy (Gen. 3:15).

There are several problems with Rolleston's connection of Orion with Jesus Christ. First, there is no legend that Orion died from a snake bite. One story was that he was stung by a scorpion, but the story does not specify that Orion was stung on his heel. Furthermore, a scorpion is not a snake. Another story is that while Orion was swimming away after battle with the scorpion, Apollo tricked Artemis into shooting the dark object in the water (Orion's head) with an arrow. Only later did Artemis sorrowfully learn that she had killed Orion. This attempt to connect Orion with Genesis 3:15 is a tremendous stretch.

A second problem with Rolleston's interpretation is interpreting the constellation below Orion as a snake rather than a hare, as the majority opinion claims. How did she know which was the true primordial constellation? She did not; she merely chose the one that matched her thesis.

A third problem with Rolleston's interpretation is k^esîl, the Hebrew word used for Orion. Elsewhere this word is translated "fool." For instance, the eight times that the word *fool* appears in Proverbs 26, this is the word used. This particular Hebrew word carries the common connotation of the word *fool* today, someone that is particularly stupid. Thus, by the Hebrew name for him, we can see that Orion is not an individual worthy of respect and devotion. To equate this fool with a type of Christ at the very least seriously borders on blasphemy, and most Christians ought to find this offensive.

If Rolleston had been as proficient in Hebrew as required to do the word studies that she supposedly did, then she ought to have known that the Hebrew word for Orion is the same word as that for a "fool." Instead, Rolleston claimed

---

[22] Frances Rolleston, *Mazzaroth*, part 2, p. 30.

that *kᵉsil* means "bound together,"[23] while Bullinger claims that *kᵉsil* refers to a great man,[24] but these claims are patently false. Rolleston either was not qualified to do these studies, or she intentionally ignored this blasphemous connection. Bullinger, Seiss, and others ought to have known better than this, but, alas, they did not, or they chose to go with their pet thesis instead. This is an example of gospel in the stars proponents ignoring biblical names for stars, opting instead for pagan sources, because those sources allegedly support their thesis. However, this clearly is inconsistent with their assumptions.

Star Names and Meanings According to the Gospel in the Stars

Rolleston advanced the notion that the original meanings of words could be found by treating them as homophones of Hebrew words or related Semitic roots. As we saw earlier, supporters of the gospel in the stars argue that the word decan comes from a Hebrew word meaning "pieces," when in reality it comes from the Latin and Greek root meaning "ten." Similarly, advocates of the gospel in the stars claim that the word *zodiac* does not mean "circle of animals," but instead means "a way having steps." Many stars have proper names, and we can trace the meanings of many of those names, often in Arabic, but sometimes in other languages, such as Greek or Latin. However, using her method, Rolleston claimed to find meanings in Hebrew or other Semitic languages for many star names that were relevant to her thesis.

Origin of Star Names

As already mentioned, the most significant source of ancient astronomical information is the *Almagest*. Books 7 and 8 of the *Almagest* contain a catalog of 1,022 stars. For each of the stars in his catalog, Ptolemy recorded the magnitude, ecliptic longitude and latitude, and also described the location within its respective constellation (descriptions such as "the shoulder of the centaur"). The magnitudes and ecliptic coordinates permit us to unambiguously identify most of the stars. The descriptions of the locations of the stars within the constellations allowed much later artists and cartographers to produce figures of the constellations on atlases, planispheres, and star globes. One of the most beautiful and best known of these is Johannes Bayer's *Uranometria* in 1603.

After the Muslim conquest, the Arabs widely used use Ptolemy's *Syntaxis*, which quickly became known as the *Almagest*. The first translations of the

---

[23] Ibid, part 2, p. 10.

[24] E. W. Bullinger, *The Witness of the Stars*, 125.

Almagest into Arabic were in the ninth century. Most of the star names we have today are Arabic, and probably date from this period when Ptolemy was popular in the Arab world. By the 12th century, the *Almagest* began to reappear in the West. The *Alphonsine Tables*, a book that proscribed how to compute the positions of the sun, moon, and planets in the Ptolemaic model was produced in Spain. This work drew heavily upon the *Almagest*. These tables originally were written in Spanish, which is very interesting, because it was highly unusual for a scholarly work to be published in a common language at that time. Eventually the tables were translated into Latin. The *Alphonsine Tables* were popular in the West for three centuries, but were eventually abandoned with the adoption of the heliocentric model.

The *Almagest* contains the proper names of only five stars, so where did the other star names come from? The star names that have been handed down to us are a mishmash of different derivations. A few star names are from ancient Greek and Roman names, and some Latin names arose in the medieval period. Most of the names are Arabic, with many coming from corruptions of the Arabic in the *Alphonsine Tables* in Spanish or Latin versions. These are often transliterations from Arabic of Ptolemy's descriptions of stars' locations within their respective constellations. A few names are of fairly recent origin.

Since many of the names have undergone translation and transliteration, there are wide variations in spelling, and there is uncertainty as to the origin and meanings of some names. A number of attempts to find the origin and meaning of star names began about 1600. One of the more exhaustive books was *Untersuchungen über den Ursprung und die Bedeutung der Sternnamen* (*Investigations on the Origin and Significance of the Names of Stars*) by Ludewig Ideler in 1809. This book, in German, remained a classic source for nearly 150 years, including the time during which Rolleston was engaged in her research. In 1882 W. H. Higgins published the short book *The Names of the Stars and Constellations*, largely relying upon Ideler. Higgins' book, like Rolleston's, amounted mostly to notes, and the author intended to expand this work with a later book but never did.

In 1899 Richard Hinckley Allen published his definitive book, *Star Names and Their Meanings*. Allen prepared a revision some years later, but this revision was never published. Instead, the 1936 edition cleaned up many of the typographical errors of the original edition. This book has remained in print since a 1963 Dover edition, and it has come to be viewed as the authoritative source on the meanings of stars' names. However, in recent years, the book has come under some criticism.

One major concern is that Allen was not a scholar of Arabic, and as such, he relied heavily and uncritically upon Ideler. But Ideler did not have access to the best Arabic sources. Most earlier historians of astronomy had endorsed Allen's assessment that the Arabic names were nothing more than transliterations of Ptolemy's descriptions of the locations of stars within their respective constellations. However, now historians think that at least a few star names that the Arabs already had prior to contact with Ptolemy's work may have been overlain in the *Almagest*. In recent years, Paul Kunitzsch, who is an Arabic scholar, has researched the history and origin of star names, and his work currently is considered the best on the subject. The most available source for Kunitzsch's work is the brief 2006 book *A Dictionary of Modern Star Names* by Kunitzsch and Smart.

The Judge

As an example of the manner in which supporters of the gospel in the stars determine the meanings of star names, let us begin with Deneb. Deneb, also known as Alpha Cyngi, is the brightest star in the constellation Cygnus, or the Swan. Deneb marks the tail of the Swan. Deneb means "tail" even in modern Arabic. This is the shortened version, for it derives from *dhanab al-dajaja*, which means "the hen's tail." *Hen* is the term for a female swan. Hence, the standard interpretation of this star's name is very reasonable.

According to the gospel in the stars thesis, Deneb means "the judge or lord who cometh quickly." Supposedly this comes from the Hebrew word דִּין (*dîn*), which means "judge," and the verbal root בהל (*bhl*) for "cometh quickly." The word *dîn* indeed means "judge," but it is a verb, not a noun. Furthermore, *bhl* actually means "to be terrified" and only derivatively, "to make haste." The 24 times that it appears in the Old Testament, it means, "to judge" or a related action. A better Hebrew word for a judge, as a noun, is דַּיָּן (*dayyān*). This is the word used for the Old Testament judges. The incorrect use of a verb as a noun seriously undermines confidence in the supposed meaning of this star's name.

There are two other star names with the Arabic word for tail. One is Denebola (Beta Leonis). Gospel in the stars advocates see this as meaning "the judge swiftly coming," which is not significantly different from their meaning claimed for Deneb. An alternate name for Denebola is Al Defera, a name that originated from the *Alphonsine Tables*. Rolleston says that this name means "the enemy put down/thrust down." The intended Hebrew root is נדף (*ndp̄*), but it actually means "to scatter, destroy." However, Al Defera comes from an Arabic word for the tuft at the end of the tail of a lion.

The third star with "deneb" as part of its name is Deneb Algiedi (Delta Capricorni). This name is a transliteration from the *Almagest* and means "tail of the goat," Capricornus being a goat. *Deneb* is Arabic for "tail," *al* is an Arabic article, and *giedi* is the Arabic word for "goat." There is even a related Hebrew word (*gᵉdî*) which refers to the kid of either a goat or a sheep. There is some further garbling of these meanings within the gospel in the stars thesis. Rolleston deleted the last part of this star's name, repeating the name and meaning for Deneb in Cygnus. But Bullinger, apparently noticing that *al gedi* means "the goat," and noting that a goat was a sacrificial animal, rendered this as "the sacrifice cometh." However, the Hebrew definite article is *ha* (ה־), not *al* (אל־); perhaps Bullinger was thinking of Arabic, in which the definite article is *al*.

Rolleston also saw the Hebrew word for "judge" in the name of an entire constellation—Eridanus, or the river. She said that the name meant "river of the judge or ruler."[25] Rolleston saw three Hebrew roots here. The first two syllables were to come from יְאוֹר (*yᵉʾōr*), meaning "stream." However, only in late biblical Hebrew (and rarely, at that) does this word mean "stream." It is used in most books as a proper reference to the Nile River or the canals that were offshoots thereof. The third syllable again comes from the Hebrew verb, "to judge." Rolleston's intended Hebrew word for the final syllable is קֹרֵס (*qorēs*), a participle meaning to "bend (over)." It is not at all clear what role this latter word played, other than supplying the final syllable, but the first syllable of that word is missing. The name Eridanus is not what the Greeks called this constellation; they simply called it *potamos*, the Greek word for "river." In Greek mythology, *Eridanos* was the name of an unspecified river somewhere in central Europe. Many eventually came to associate *Eridanos* with the Po River of northern Italy. "Eridanus" is the Latinized version of *Eridanos*. It appears that this Latin name for the river came into use during the Medieval Period, as did most other Latin names for constellations. Overall, Rolleston's meaning for Eridanus is untenable, and even if her derivation were true, it is not at all clear how this would relate to the gospel message.

---

[25] Frances Rolleston, *Mazzaroth*, part 2, p. 11.

Some Embarrassing Examples

The two brightest stars in the constellation Delphinus, or the dolphin, are Svalocin and Rotanev, which Rolleston rendered "Scalooin" and "Rotaneb." These two star names are not ancient, but instead began appearing on star charts in 1814. Many years ago it was discovered that these two star names are "Nicolaus Venator" spelled backwards. Venator, a Russian, was an assistant to the great Italian astronomer Giuseppe Piazzi, and is often referred to by the Latin equivalent, Niccolo Cacciatore. These two star names seem to have surfaced in Italy during their lifetimes, though it is not known who gave the stars these names. However, Rolleston somehow managed to find meanings for these names in Arabic, Syriac, and Chaldean, and others have uncritically repeated her blunder.

Rolleston lists the Hebrew meaning for Mira, the famous variable star in the constellation Cetus, as "the rebel."[26] The intended Hebrew root is מרה (mrh). There are two problems here. First, this word is a verb meaning "to be rebellious" rather than a noun. Second, Mira was not listed by any ancient sources, and as such this name is of relatively recent origin. Credit for the discovery of the variable nature of Mira goes to David Fabricius in 1596. In 1662 Johannes Hevelius named the star "Mira" in his *Historiola Mirae Stellae*. The name means "wonderful," or "astonishing," and comes from the Latin word *mirus* (we get the word *miracle* from this word). How could Rolleston have found an ancient Hebrew meaning in a star name that originated only two centuries before the publication of her book?

The Names of First Magnitude Stars

Rolleston stated that the first magnitude star Aldebaran, the brightest star in Taurus, means "the leader."[27] Here Rolleston's intended Aramaic word is *hadābar*. Actually, the name Aldebaran comes from Arabic for "the follower," presumably because it follows behind the Pleiades as the earth rotates. Therefore, Rolleston found exactly the opposite meaning for this star's name.

On the same page, Rolleston claims that Betelgeuse, the brightest star in Orion, means "coming." It is generally agreed that the name means "armpit of the central one." This is appropriate, because Betelgeuse does mark the armpit, or shoulder, of Orion.

---

[26] Ibid, part 2, p. 9.
[27] Ibid, part 2, p. 10.

Rolleston says that Capella, the first magnitude star in Auriga, comes from Latin and means "the goat, atonement."[28] She is partly correct, for the name comes from the diminutive of the feminine word for goat, so this literally is "the little she-goat." There are at least two problems here. First, Rolleston made an obvious connection to a sacrificial goat, but the atonement sacrificial goat was to be a male, not female (Lev. 1). Second, she admits to getting this meaning from Latin, but one of her assumptions was that Hebrew was the mother tongue of all. However, this name is not even close to the Hebrew or Arabic words for goat. Therefore, this meaning is not relevant if one follows Rolleston's stated methodology. Incidentally, Bullinger acknowledges the Latin origin of Capella, but he implies that Alioth was the original name for this star, though there is no evidence for this.[29] Apparently, Bullinger was so convinced that the Latin name was a translation of the original star name, he simply asserted that this was the case.

Rolleston stated that the Hebrew meaning of the star Regulus is "the treading under foot."[30] Actually, the name is the Latin diminutive form of "king." The word "regal" comes from a similar root. Regulus is the brightest star in the constellation Leo, or the lion. We usually think of a lion as being a royal beast, so the name fits. Rolleston had the Hebrew word for foot, רֶגֶל (regel), in mind here.

A similar Arabic word does lend its name to another star, Rigel, the second brightest star in the constellation Orion. Rolleston found the meaning "the foot, or who treadeth under foot" for this star,[31] which obviously is an appeal to Genesis 3:15. Rolleston got part of this right, for Rigel comes from the first word of *Rijl Jauzah al Yusra*, meaning "the left leg of Jauzah," "Jauzah" being an early Arabic name for Orion. However, this name as now known first showed up in the *Alphonsine Tables* in 1521, so there is some question about Rolleston's derivation, and her second meaning reads far too much into the word.

Rolleston claimed that the name of the bright star in Scorpius, Antares, comes from Arabic and means "the wounding."[32] Rolleston's derivation is unclear here. She gave the meaning as "the wounding (Arabic form) (cutting)," and she gives Jeremiah 36:23 as the reference. Apparently, the Hebrew root

---

[28] Ibid, part 2, p. 11.

[29] E. W. Bullinger, *The Witness of the Stars*, 134.

[30] Frances Rolleston, *Mazzaroth*, part 2, p. 15.

[31] Ibid, part 2, p. 10.

[32] Ibid, part 2, p. 19.

intended is קרע (*qrʻ*), but how this morphed into Antares is a mystery. Actually, Antares is a transliteration of the Greek word used by Ptolemy, meaning "like or in the place of Mars" (anti-Ares, Ares being the Greek equivalent of the Roman God of war). Antares has this name because in its brightness and color it often resembles Mars.

Problems with Ursa Major

There are several problems with how supporters of the gospel in the stars handle Ursa Major, or the Big Bear. Rolleston says that the Hebrew name for the constellation is "Ash," meaning "the assembled."[33] From this word, supporters of the gospel in the stars claim that this constellation originally was a sheepfold, not a bear. However, there is no evidence that this constellation ever was a sheepfold. The word עַיִשׁ (*ʻayiš*) appears in Job 9:9 and Job 38:32, but there is no evidence that this word comes from a root meaning "to assemble oneself," as supporters of the gospel in the stars claim. This word has been translated various ways. The Septuagint translated it "Arcturus" in Job 9:9, but rendered it as "Evening Star" (presumably Venus) in Job 38:32. Jerome translated it as Arcturus in either verse, which the King James Version followed. The Revised Standard Version, New American Standard Bible, New International Version, and the English Standard Version all translate the word as "bear" in these two verses.

Arcturus is the name of a bright star in the constellation Boötes, but these two verses almost certainly do not refer to this star. Boötes is one of the few Greek constellation names; it is a transliteration of the Greek word for wagon driver. "Arcturus" comes from *arktos* and *ouros*, the Greek words for "bear" and "guard," so the name means "guardian of the bears." This name probably comes from the close proximity of Arcturus to Ursa Major and Ursa Minor. Incidentally, Rolleston claimed that the meaning of the Greek word *arctos* is "traveling," from the Hebrew word אֹרְחָה ("traveling company, caravan").

So what is this object mentioned in Job 9:9 and Job 38:32? Several possibilities have been proposed. Besides Arcturus, the bright stars Capella and Aldebaran have been suggested. Another suggestion is that it is the Hyades star cluster. This makes sense, because the Hyades is located close to Orion and the Pleiades, the other two astronomical things mentioned in Job 9:9 and Job 38:32. However, the vast majority of opinion is that it refers to

---

[33] Ibid, part 2, p. 13.

the Big Bear. How is this opinion reached? The Hebrew word 'ayiš is similar to the Arabic word for a bier, or funeral platform. The Arabs long referred to the square in the Big Dipper (a part of Ursa Major) as a bier. Hence, this could be the origin of the Hebrew word used in Job 9:9 and Job 38:32. That is, the Hebrew word used here may have been borrowed from another Semitic language.

Apparently, Rolleston accepted the majority opinion that the verses in Job refers to the Big Bear, from which she inferred that the Hebrew word used in these two verses was the ancient Hebrew name for the constellation. However, the only Hebrew sources that we have on the constellations say that their name for the constellation is דֹב (dōḇ), which is the Hebrew word for "bear," as found in 1 Samuel 17:34, 36–37. Admittedly, these sources are medieval, and not ancient.

This constellation has almost universally been known as a bear, even among Native Americans. This coincidence argues for a very ancient origin for the Big Bear. Rolleston's claim that the ancients knew this constellation as a sheepfold is without foundation and amounts to an assertion on her part. Part of the reasoning for making a sheepfold out of the bear may have come from Dubhe, the name of one of the stars in Ursa Major. Rolleston claims that this means "a herd of animals," and Bullinger concludes that the likely intended animals are sheep. Rolleston and followers argue that the name derives from the Hebrew word דֹּבֶר (dōḇer), which is translated "fold" in Micah 2:12. However, the Hebrew word for "bear," dōḇ, is a much better fit, and the Arabic word for bear is very similar. There is another problem with Rolleston's claim, for dōḇ has two Hebrew consonants (called daleth and beth), which are the same consonants in Dubhe, whereas dōḇer has these consonants plus the Hebrew consonant resh.

The star Beta Ursae Majoris is called Merak. Rolleston says that this is from the Hebrew word עֵדֶר ('ēder), meaning "flock (of animals)," or the word derives from an Arabic word that means "purchased." Well, which is it? Neither. The name actually means the "loin (of the bear)" in Arabic.

Gamma Ursae Majoris is called Phecda, which comes from the Arabic for "thigh," for this star marks the thigh of the Great Bear. Rolleston says that the name comes from a Hebrew root that means "visited, guarded, numbered." The Hebrew root that she had in mind is פקד (pqd). Notice that transforming pqd into Phecda requires shuffling some of the letters.

Delta Ursae Majoris is Megrez, which comes from the Arabic for "root of the tail," for it is located at the base of the tail of the Great Bear. Rolleston wrote that the name means "separated, as the flock in the fold, cut off." Her intended Hebrew root here is גרז (*grz*), which again requires some shuffling of letters to get to "Megrez."

Bullinger gives the meaning of Epsilon Ursae Majoris, Alioth, as "the she-goat, or ewe."[34] The intended Hebrew word is עֵז (*ēz*). This is not a good fit, for it requires additional consonants.

Rolleston misidentified Zeta Ursae Majoris as Epsilon Ursae Majoris (obviously a misprint), though Bullinger corrected this error in his book. Both said that this star's name, Mizar, means "separate," with the intended Hebrew root being נזר (*nzr*). However, the Hebrew root *nzr* carries the sense of something like "to consecrate oneself to deity." However, Joseph Justus Scaliger (1540–1609) had improperly changed the name from Mirak to Mizar, Arabic for girdle or waistcoat. Therefore, the Hebrew word claimed by advocates of the gospel in the stars as the correct meaning is not correct. Close to Mizar appears the fainter star Alcor. Rolleston wrote that the name is Arabic for "the lamb." The intended Hebrew word of origin is כַּר (*kar*). However, this word refers to a ram, not a lamb. Furthermore, most sources say that the name comes from the Arabic for "the faint one." This works very well, for Alcor is much fainter than the nearby Mizar.

A Few Other General Examples

Zuben el Chamali and Zuben el Genubi are two of the brighter stars in Libra, or the scales. These two names are Arabic for "northern claw" and "southern claw" respectively, because some cultures considered these stars in Libra to be the claws of Scorpius the scorpion, which is a nearby constellation. Advocates of the gospel in the stars claim that these two names mean "the price which covers" and "the purchase" or "the price which is deficient."

The proper name for the star Eta Geminorum is Propus. This name is the transliteration of the Greek word for "foot," for this star is the left foot of Castor, one of the Gemini twins, in the description of Ptolemy. This name began appearing as the transliteration from Ptolemy during the Renaissance. Rolleston gives the meaning from Hebrew as "the branch, spreading." The

---

[34] E. W. Bullinger, *The Witness of the Stars*, 155.

intended Hebrew words are פֹּארָה (pu'rāh), meaning "bough" and פּוּשׁ (pûš) meaning "spread."[35] Rolleston's error is thus quite an embarrassing one.

Conclusion

There are many other examples that we could list. Suffice it to say that the overwhelming majority of the meanings that supporters of the gospel in the stars give for words and names of stars are at complete variance with other, more reliable sources. They are correct in a few instances, but most of the correct meanings have little, if anything, to do with the thesis about the gospel message being preserved in star names.

One example of a correct meaning given by Rolleston concerns Fomalhaut, the brightest star in Piscis Australis, or the Southern Fish. Fomalhaut comes from Arabic, meaning "the fish's mouth." Rolleston did not embellish upon what possible soteriological meaning Fomalhaut or Piscis Australis might have, so it is not clear what possible gospel-related meaning she saw in either of them. Seiss did not even mention Fomalhaut, but he did speculate on some possible meanings for the constellation. He began his speculation with these words:

> The mythic legends do not help us much with regard to the interpretation of this constellation, but they still furnish a few significant hints.[36]

After mentioning some pagan legends about Piscis Australis, Seiss inferred connections to the church as the bride of Christ. Bullinger, following Rolleston, did give the correct meaning for Fomalhaut, but in his more terse style, he abruptly and incredibly concluded about the constellation,

> It sets forth the simple truth that the blessings procured by the MAN—the coming Seed of the woman, will be surely bestowed and received by those for whom they are intended. There will be no failure in their communication, or in their reception. What has been purchased shall be secured and possessed.[37]

Though obliquely related—Bullinger concentrated on Christ while Seiss emphasized the church—these meanings are very different. To her credit, Rolleston apparently could see no connection to the gospel in Piscis Australis, so she quietly let this group of stars go by. But Seiss and Bullinger, wrapped

[35] Frances Rolleston, *Mazzaroth*, part 2, p. 12.
[36] Joseph A. Seiss, *The Gospel in the Stars*, 75–76.
[37] E. W. Bullinger, *The Witness of the Stars*, 89.

up in enthusiasm for the gospel in the stars thesis, blundered on with odd speculations. More recent treatments of the gospel in the stars have further embellished this nonsense.

The fact that supporters of the gospel in the stars theory see so many different things in constellations and star names shows that this amounts to a Rorschach test—one sees what one wants to see in them. If the gospel in the stars theory were correct, there would be more agreement as to meanings of constellation and star names. The sheer volume of the incorrect meanings ought to be an embarrassment for those who subscribe to the gospel in the stars theory and certainly argues strongly against the correctness of their thesis. In the next chapter we shall explore further problems with the gospel in the stars theory.

CHAPTER 16

The Gospel in the Stars: Additional Problems and Conclusions

The most influential books on the gospel in the stars probably are those by Joseph A. Seiss and E. W. Bullinger. Neither Bullinger nor Seiss included much in the way of references or even allusions to original sources. Rather, they merely declared the meanings of various star names and constellations. From the text of their books, no one could judge where the material originated or how they derived the meanings of names. Both gentlemen derived their work solely from Frances Rolleston, for they both gave the credit to Rolleston in their books' prefaces. Indeed, they commended her for her diligent work in searching old texts and deciphering the meanings of names in the original languages.

Today's defenders of the gospel in the stars claim that Rolleston studied many ancient sources to find her information. To her defenders, the use of ancient sources adds tremendous weight to the argument for the gospel in the stars. Indeed, if the proper sort of scholarship were applied to original sources, then this would add weight to the case. However, how good was Rolleston's research? As it turns out, it was not very good.

Poor Scholarship

Rolleston's scholarship does not favorably meet up to modern standards. One problem is that she cites very few of the works that she used. For example, for the Hebrew names of constellations and stars, Rolleston lists the Hebrew sources of "Buxtorf's Rabbinical Lexicon, etc."[1] This rabbi evidently was Johannes

---

[1] Frances Rolleston, *Mazzaroth* (1875; repr., York Beach, ME: Weiser Books, 2001), part 2, pp. 11, 14.

Buxtorf the Elder (1564–1629), and if so, then the book's full title was *Lexicon Hebraicum et Chaldaicum cum brevi Lexico Rabbinico Philosophico* published in 1607. The "etc." must refer to other unnamed lexicons. For Syriac, Rolleston listed "Hyde's Syntagma and Comment, etc." The author must be Thomas Hyde (1636–1703), and his book must be *Syntagma dissertationum quas olim Thomas Hyde separatim edidit*, a collection of unpublished Hyde manuscripts assembled by Gregory Sharpe and published in 1767. For Greek constellation names she listed "Aratus, Ptolemy, etc.," and for Latin constellation names she listed "Cicero, Virgil, Ovid, etc." As before, the "etc." must refer to other unnamed sources. Other sources were listed by abbreviated names, and hence they are difficult to identify. It appears that most, if not all, of these sources were in Latin.

Another example concerns her reference to "Albumazer." This is Abu Ma'shar Al-Balkhi (787–886), a leading Persian astrologer. Rolleston incorrectly stated that he was "the great Arab of the caliphs of Granada."[2] Later writers on the gospel in the stars repeated this incorrect information. Albumasar (the preferred spelling today) wrote several books, all of them more about astrology than astronomy. Most of these works eventually were translated into Latin and used in the west during the Middle Ages. Although Latin versions exist in a few very exhaustive library collections, most of these works have not been translated into English or any other modern language. Since Rolleston and later advocates of the gospel in the stars quote or reference Albumasar without a specific work being identified, it is very difficult to check references to Albumasar.

A second problem with Rolleston's scholarship is that even when a particular book is cited, she rarely indicated the location within the work to support her claim. A third concern is that there is no good reason to believe that Rolleston read any of her sources in the original languages or even checked her claims with scholars competent in those languages. We do not know much about the life of Rolleston. Was she educated in ancient and Middle Eastern languages? We do not know, though it is unlikely that she was. Every one of Rolleston's sources that she clearly identified was available in a Latin translation in her day. If she were formally educated at the time, she might have been proficient in Latin. On the other hand, her creation of the decans as discussed in Chapter 15 may indicate that her ability to read Latin was poor.

[2] Ibid, part 5, p. 15.

A Recent Idea

The primary claim of the gospel in the stars thesis is that the gospel story in the stars was known to the patriarchs before the Flood, but in time was forgotten. It is common to claim, for example, that God had to reintroduce the concept to Abraham because it had already been forgotten. Even the gospel in the stars advocates who claim it was revived at the time of Abraham believe it was lost again by the time of Moses, for that was why God finally inspired the writing of Scripture.

This means the gospel in the stars was forgotten by the 15th century BC, and possibly as early as 1,000 years before. Unfortunately, anything approaching complete manuscripts from antiquity is exceedingly rare. A thousand-year-old manuscript is very old. For instance, the writings of such greats as Aristotle often date more than a millennium after their deaths. We do not even have translations or copies of books by known authors before about the eighth century BC. If the gospel in the stars was antediluvian as claimed, then there were two millennia to garble the message before any sources that we have regarding the names of stars and constellations were written.

Even if the message had been kept clean by a remnant through Abraham and down to Moses, there still is a gap of a thousand years. Not only are there no texts preserving the original knowledge of the gospel in the stars, but we know of no sources before Rolleston that claimed there ever were such texts, or that anyone ever lived who believed such things. The first source we know of to make this claim is Rolleston's book.

Hence, it is obvious from Rolleston's book that she had no texts that clearly underpinned her proposal. Instead, she created the meanings of star and constellation names to support her theory. The evidence we have is most consistent with the gospel in the stars thesis not being an ancient idea at all, but entirely the invention of Frances Rolleston less than 200 years ago.

In Chapter 15, I mentioned 19th-century writings on constellation and star names. There are other more recent writers who, while they did not write works exclusively dedicated to star names, did touch upon the subject. In 1877, Richard A. Proctor wrote *Myths and Marvels of Astronomy*, which contains some discussion on the origin of the constellations. Proctor was a proficient writer on popular astronomy, as was his daughter, Mary Proctor, who also wrote some on star lore. In 1903 the Italian astronomer Giovanni Schiaparelli wrote a well-researched book, *Astronomy in the Old Testament*. This was in

Italian, but an English translation followed in 1905. The famous astronomer E. Walter Maunder wrote *Astronomy of the Bible: An Elementary Commentary on the Astronomical References in the Holy Scripture* in 1908, where he discussed biblical references to astronomical bodies. All of these gentlemen were well educated in the Bible and obviously took the Bible very seriously. None of them saw anything resembling a gospel in the stars.

Furthermore, several learned people of the past were quite alarmed with the obvious pagan roots of the constellations and attempted to change the situation. In his 1627 book, *Coelum Stellatum Christianum*, Schiller made some direct correspondences with biblical themes, such as Ara, or altar, being replaced by the alter in the Tabernacle, Argo Navis being replaced by Noah's Ark, and Columba, or the Dove, being replaced by the dove sent out by Noah. Supporters of the gospel in the stars make much of Virgo, Taurus, and Aries, but both Schiller and much earlier the noted early medieval scholar Bede totally passed on this obvious comparison to biblical themes. The fact that Schiller and others saw no parallel between the Bible and these zodiacal signs is very interesting, if the gospel in the stars theory has any merit. Bede and Schiller had a similar concern to those who support the gospel in the stars—a concern about the paganism present in the constellations. And being much earlier in time, both Schiller and Bede might have had access to earlier manuscripts. Their solution was not to attempt to reclaim original biblical truth, for they did not see this in the constellations. Rather, they sought to remove and replace the paganism with Christian meaning or signification of their own making.

False Antiquity Assignments

Several of Rolleston's sources were not derived from the ancient wisdom she assumed they were. For example, one of Rolleston's most important medieval sources is Rabbi Avraham Ben Meir Ibn Ezra (1092/93–1167). Note that different authors identify this man by various spellings and combinations of his titles and names—Rolleston used Aben Ezra. Aben Ezra was a Jew from the Iberian Peninsula, but who traveled extensively in Europe, North Africa, and the Middle East. He is known as a prolific poet, Jewish commentator, and writer on various subjects such as math, science, and astrology. It is important to know that he wrote his works in Hebrew, and many of his science writings were translations of Arab manuscripts available in Moorish Spain and North Africa.

Presumably Aben Ezra translated at least portions of the *Almagest* and Arabic astronomical lore into Hebrew. He was popular with his readers, because his Jewish audiences generally were ignorant of these topics. This is very significant, because rather than informing us about ancient Jewish astronomical lore, he may have *introduced* astronomical lore to medieval Jewish people. This is an important distinction, because Rolleston apparently believed that Aben Ezra knew much about ancient Jewish astronomy, when in reality it is extremely doubtful that he knew about it at all. In other words, while Rolleston assumed that Aben Ezra was a source of ancient Hebrew astronomical lore, he actually was a conduit of astronomical lore from ancient Gentile sources to medieval Jews.

A second example is Ulugh Beg (1393/94–1449), a Timurid leader, mathematician, and astronomer. Of Mongol descent and born in Persia, Ulugh Beg spent most of his life in Samarkand, where he built an observatory. Perhaps his greatest contribution was an updating and correcting of the *Almagest*. His corrections consisted mostly of remeasuring stellar positions at his observatory, but he also included updated Arabic names for many stars. His catalog of nearly a thousand stars was the first catalog since Ptolemy's.

There are several important points about his work. First, being Muslim, he wrote and worked with Arabic sources, primarily the *Almagest*. Second, he contributed new observations, but he did very little to inform us on ancient star lore. He lived more than a millennium after Ptolemy, so one would expect Ptolemy to be much closer to ancient sources than Ulugh Beg was. For a long time, most scholars thought that Ulugh Beg transmitted no Arabic astronomical lore, but of late, scholars tend to think that he did further some Arabic lore prior to the Muslim conquest of the Middle East in the seventh century, though it is difficult to discern exactly what this content would have been. Nor is it clear that anything he might have added predated Ptolemy's time, rather than being from the later Christian era. In fact, there is no evidence that Ulugh Beg had access to any ancient sources that pre-dated Ptolemy. Rolleston may have assumed too much and thus may not have properly assessed the work of Ulugh Beg. Consider this statement:

> Ulugh Beigh…who lived about the middle of the fifteenth century, is considered to have transmitted the ancient Arabian science.[3]

---

[3] Ibid, part 2, p. 14.

Rolleston does not define what she means by "ancient Arabian science," but it easily could imply to her readers lore predating the Christian era. If true, then Ulugh Beg's writing would have been very significant in deciphering ancient meanings. But, alas, it is unlikely that Ulugh Beg's writings contributed anything at all from the pre-Christian era. Instead, he corrected the earlier Arab translation of the *Almagest*, done by several scholars, the most notable being 'Abd Al-Rahman Al Sufi (903–986). Thus, since both Aben Ezra and Ulugh Beg derived their work from Ptolemy, it is unlikely that any truly ancient (pre-Christian and pre-Ptolemy) lore is found here.

Al Sufi is credited with including with his translation of Ptolemy the Arabic names of stars from the time prior to western influence. However, the western influence arrived about the time that Al Sufi did his translation, so earlier Arabic names may only go back a few centuries. To be truly ancient and significant in the way that Rolleston meant would require that his Arabic names go back two millennia prior, which is extremely doubtful. Rolleston's assumption that Ulegh Beg's Arab names are from antiquity is not supported by the views of any scholars of this and related works except in a few specific instances.

Modern Scholarship on the Origin of Constellations

In modern scholarship there is no consensus as to who originated the constellations. The dominant belief is that they originated with the ancient Babylonians (not the neo-Babylonian empire). From the Babylonians the constellations were transmitted to the Egyptians, and the Egyptians in turn passed them on to the Ancient Greeks, though there may have been some more direct transmission from the Babylonians to the Greeks. Shortly thereafter, the Romans absorbed much of the constellation lore from the Greeks. A popular variation on this history is that the ancient Minoan civilization originated the constellations and passed them on to the Babylonians.

Whereas most of the star names are Arabic, most of the constellation names are Latin. The 40 or so relatively modern constellations bear Latin names, for Latin had been the preferred language of science for some time when many of those were named. However, even most of the 47 remaining of Ptolemy's 48 original constellations bear Latin names, usually Latin translations of the Greek words that Ptolemy used. For instance, "Ursa Major" and "Ursa Minor" are Latin for the Large and Small Bears, and "Canis Major" and "Canis Minor" are the

Large and Small Dogs." This is true of the zodiacal signs as well—"Leo" is Latin for lion, and so forth. We often say "Virgo, the Virgin," or "Cygnus, the Swan," but this is redundant. More correctly, we ought to say "Virgo," or "the Virgin."

Problems with the Zodiac in the Bible

Rolleston found many zodiacal connections in the Bible. For instance, she claims that there are allusions to the 12 signs of the zodiac in Jacob's blessing on his 12 sons in Genesis 49.[4] Rolleston also claimed the same correspondence between the tribes of Israel and the signs of the zodiac is contained in the final blessing of Moses in Deuteronomy 33.[5] There are at least two problems with this. First, one is hard pressed to find any commentary on Genesis or Deuteronomy to support this interpretation. Second, the alleged connections between the zodiacal signs and the blessings for each son or tribe are extremely creative. The only obvious, possible connection is the one supposedly between Leo and Judah in that Judah is referred to as a lion. However, there is not even a hint in the biblical text of the other 11 connections that Rolleston made. Rolleston further speculated that each son carried a symbol from one of the signs of the zodiac and that Jacob pointed to each zodiacal symbol in pronouncing his blessing. This is entirely speculation on Rolleston's part.

Building on this, Rolleston also claimed that in the wilderness each of the tribes carried a standard with the tribe's respective zodiacal sign inscribed upon it.[6] It is true that the tribes had standards, or banners (Num. 2:1–34; 10:11–28), but Scripture does not record what was on the standards. Rolleston credited Josephus with this information.[7] However, there is no mention of this in either Josephus's *Antiquities of the Jews* or *Wars of the Jews*. Rolleston offered no other references for any of this. Sadly, Seiss and Bullinger uncritically repeated this claim concerning the zodiacal connection of each tribe, and both fully endorsed the specious claim that the tribal standards in the wilderness wandering were zodiacal. For instance, Bullinger confidently stated,

And it is more than probable that each of the Twelve Tribes bore one of them (a zodiacal sign) on its standard.[8]

---

[4] Ibid, part 1, p. 17.
[5] Ibid, part 2, p. 38.
[6] Ibid, part 1, p. 12; part 2, p. 48.
[7] Ibid, part 2, p. 48.
[8] E. W. Bullinger, *The Witness of the Stars* (1893; repr., Grand Rapids, MI: Kregel Publications, 1967), 17.

Rolleston further claimed that each of the 12 stones on the breastplate of the high priest contained an inscription of the zodiacal sign of the respective tribe.[9] However, this contradicts the clear text of Exodus 28, which informs us that each stone bore the name of a tribe. Where did Rolleston get this? Josephus (*Antiquities of the Jews*, 3.7.5) describes the breastplate similarly to how it is described in Exodus 28. However, two sections later Josephus speculated that each of the stones may correspond to the months of the year or to the zodiacal signs, though it is not clear if he is offering his own opinion or the opinion of Jews of his day. The introductory passage to this section makes it clear that Josephus was attempting to answer questions from secular sources. Rolleston may have misunderstood what Josephus wrote.

It is likely Rolleston's erroneous claim that Josephus stated the banners of the tribes in the wilderness bore zodiacal signs stemmed from a misunderstanding or further inference about the breastplate stones. Again, Josephus did not state that the stones bore zodiacal signs. Rather, he opined that the stones, being 12 in number, might have *corresponded* to the 12 zodiacal signs. Once one assumes that this implies that each stone had a zodiacal inscription, one for each tribe, it is very easy to infer that the banners must have had those same zodiacal signs on them as well. The fact that Josephus, being a Jew, did not seem to understand that the 12 stones represented the 12 tribes and opined instead that they might have referred to the 12 zodiacal signs indicates that Josephus had no real problem with astrology.

It is true that there is some Jewish tradition connecting each of the 12 tribes to a particular zodiacal sign. However, sources for these associations date very late, from the Middle Ages, which suggests that the correspondences are not ancient at all. Furthermore, there is no single system of correspondence. Instead, different sources claim different systems of assigning particular signs to each tribe. If the identification of each tribe to a unique sign were from the patriarchs or the wilderness wandering, there ought to be a unique identification of each tribe.

To further support this connection between the zodiacal signs and the 12 tribes, recently some advocates of the gospel in the stars have pointed to zodiacal floor motifs found in ruins of several synagogues dating from the fourth to sixth centuries AD. Most notable is the sixth-century Beth Alpha Synagogue in Israel. Some have gone so far as to claim that some or all of these

---

[9] Frances Rolleston, *Mazzaroth*, part 2, pp. 45–46.

zodiacal signs are explicitly identified with particular tribes in these mosaics. However this is not true, for the inscriptions are the names, in Hebrew, of the zodiac signs. For instance, the sign for Leo has the Hebrew inscription אַרְיֵה, meaning "lion."

Not only is there no connection made to the Old Testament, but at the center there is a depiction of Helios, the Greek god of the sun. One has to ask, assuming that these ruins indeed are synagogues, why such blatantly astrological and pagan depictions are found in a synagogue, when such things were forbidden to the Hebrews? It is likely that these mosaics merely were decorative art. By the sixth century, many Jews had become thoroughly Hellenized to the point that the Old Testament prohibitions against astrology and paganism were of no consequence to them. At any rate, the existence of these motifs in what are supposed to be ruins of synagogues do not make the case for the association of a zodiacal sign to each of the 12 Hebrew tribes two millennia earlier.

Rolleston also linked zodiacal connections to names in the Old Testament. She claimed a zodiacal connection to the names of the 12 sons of Jacob, as well as the first 12 patriarchs.[10] The latter is forced, as there were ten antediluvian patriarchs, but Rolleston had to include two post-Flood patriarchs to get the number to 12. She also found connections between zodiacal signs and types of the Levitical law, cherubic forms in prophecies, types of the apocalypse, and prophecies of the Messiah.[11] This sort of argument is contagious, as witnessed by Seiss' odd ramblings about the alphabet and the stars.[12] Besides the bizarre claims here, it also is not clear what the point of these connections, if real, was. Just how does this connect to the gospel in the stars? These weird, mystical speculations ought to give proponents of the gospel in the stars cause for concern.

Problems with Rolleston's Interpretation of the Star of Bethlehem

Rolleston included a section on the star of Bethlehem.[13] She stated that in about 125 BC a bright star appeared, so bright that it was visible during the day. Rolleston also said that this event induced Hipparchus to produce his star catalog about this time. This version of Hipparchus' motivation was supplied

---

[10] Ibid, part 2, pp. 32, 37.

[11] Ibid, part 2, pp. 49–61.

[12] Joseph A. Seiss, *The Gospel in the Stars* (1882; repr., Grand Rapids, MI: Kregel Publications, 1972), 23.

[13] Frances Rolleston, *Mazzaroth*, part 2, pp. 104–06.

by Pliny the Elder in his *Natural History*, but in his *Almagest*, Ptolemy said that it was Hipparchus' discovery of precession of the equinoxes that caused Hipparchus to produce his star catalog. However, there is no reason why both stories could not be true.

By Rolleston's comparison with much later events, such as the "new stars" seen in 1572 and 1604, today we would recognize the 125 BC event as a nova or supernova. Indeed, many modern astronomers think that it was a supernova, though not much credence is given to it, because all mentions of it come from much later, secondary sources. Rolleston went on to suggest, in an oblique manner as a series of questions, that this star remained bright for many years (into the second century AD), may have been in *Coma* (her alleged constellation of "the Desired"), and hence was the star that alerted the magi that the Messiah was born. It is notable that she did not clearly state these as "facts."

Rolleston also brought in legends about the star which are attributed to Zoroaster. Another legend that she suggested was that the magi saw the reflection of this bright star in the bottom of a well in Bethlehem, thus indicating that this star was directly overhead. Of course, if there were such a bright star, it could be directly overhead for only a few minutes each night. Rolleston suggests that since this occurred at about midnight on the winter solstice, this must have been when the magi arrived. Notice the endorsement of the traditional date of December 25 for Christ's birth, something that Rolleston endorses elsewhere; but almost no scholars today believe December 25 was the correct date. Rolleston did not document her sources for much of this, but did quote from an 1850 book, *The Star of the Wise Men*, by R. C. Trench. She particularly quoted Trench regarding early church fathers, such as Ignatius, concerning the appearance of this star. Both Seiss and Bullinger included similar discussions about this alleged "new star" being the star of Bethlehem.[14] From their descriptions, it is very clear that Seiss and Bullinger solely and uncritically relied upon Rolleston in this matter. In turn, many writers on the gospel in the stars today rely solely and uncritically upon Seiss's account.

How credible is this explanation for the star of Bethlehem? It is not credible at all. First, a nova or supernova is not readily visible for years. A nova is bright for a few days, but then quickly fades away. A supernova stays bright longer than a nova, but it too fades. For instance, the AD 1054 supernova that produced the

---

[14] Joseph A. Seiss, *The Gospel in the Stars*, 161–62; E. W. Bullinger, *The Witness of the Stars*, 36–39.

Crab Nebula was visible during the day for a few weeks, but it completely faded from naked-eye view after 14 months. Hence a nova or a supernova could not have been visible for more than a century. Chapter 7 further discusses this and other theories as to what the Christmas star was.

Second, it is not clear how, if the star had been visible for more than a century before the birth of Christ, this would have been an appropriate sign to the magi. Unless the magi were well over 100 years old, the star would have always been visible to them, so why would that be unusual? Also, what would have prompted them to finally make the journey to Israel when they did? Many modern supporters of the gospel in the stars think that the star was a conjunction of planets as well, but then this multiplies the number of stars that were seen.

Third, there is no astronomical record of this star. Modern historians of astronomy have pored through historical records to identify possible nova and supernova sightings from the past, but this event does not appear on any of those lists. It is difficult to know what to make of the alleged support from early Christian writers—as no references have been provided. Given the very poor manner in which Rolleston and those who followed her handled other information, it is very likely that this is another example of how they have created all sorts of new "facts" in support of their thesis. It is a shame that modern supporters of this view have not bothered to carefully check these extraordinary claims, opting instead to uncritically repeat them as established facts.

Why Not Use Hebrew Names Where They Are Known?

Rolleston found Hebrew root meanings for various common and traditional names associated with stars, and her successors uncritically followed her approach. However, a few names involving stars are found in the Bible. How do supporters of the gospel in the stars handle these? Unfortunately, not very well. In most of the cases, there is some uncertainty in how to accurately identify which names go with which astronomical objects.

We have already seen in Chapter 15 that Orion is mentioned three times in the Bible (Job 9:9; 38:31; Amos 5:8). We also have seen that the Hebrew word used in reference to Orion is כְּסִיל ($k^{e}s\hat{\imath}l$), a Hebrew word meaning "fool." However, Rolleston says that the word means "bound together," and Bullinger says that it means "a strong one, a hero, or giant."[15] But as we have previously

---

[15] Frances Rolleston, *Mazzaroth*, part 2, p. 10; E. W. Bullinger, *The Witness of the Stars*, 125.

seen, $k^e s\hat{\imath}l$ does not mean either of these. How a scholar such as Bullinger could have missed this is amazing.

The meaning of "Orion" is in doubt, but the best guess is that it comes from Akkadian for "light of heaven," referring to the sun. Supporters of the gospel in the stars see much significance in this, for they generally render it "coming forth as light." However, the premise of the gospel in the stars is that the names in the original language carried the true meanings of the names of the stars and constellations. Furthermore, they assume Hebrew is the mother tongue, so it stands to reason that one ought to look for meanings in the Hebrew word for Orion. But if one objectively looks at the Hebrew word for Orion, one finds that the meaning clearly *contradicts* the meaning that advocates of the gospel in the stars have gleaned.

One could argue that $k^e s\hat{\imath}l$ was not the original name for Orion, but there is no evidence for that. Certainly, in light of the fact that Orion is mentioned twice in Job, arguably the oldest book of the Bible, we have here the earliest name for Orion on record, seeing that the other ancient mention of Orion (Aratus) postdates Job by at least a millennium.[16] One could alternately argue that perhaps these biblical references do not refer to Orion at all. Indeed, the word $k^e s\hat{\imath}l$ in reference to a heavenly object is found a fourth time, in Isaiah 13:10, where the word is in the plural form and is translated "constellations" in the King James Version. The Revised Standard Version, English Standard Version, New American Standard Bible, and New International Version concur with the King James Version translation of $k^e s\hat{\imath}l$ in all four verses. Interestingly, the Septuagint agrees with Job 38:31, but uses "Orion" rather than "constellations" in Isaiah 13:10, and gives "Hesperus," referring to the Evening Star (Venus) instead of "Orion," in Job 9:9. The Septuagint renders $k^e s\hat{\imath}l$ as "all things" in Amos 5:8.

Therefore, in the case of Orion, it seems that the supporters of the gospel in the stars do not follow their own theory for the origin of the names of the constellations. How do other examples fare? The Pleiades star cluster appears three times in the Bible, in the same three verses where Orion appears in the King James Version. In the two verses in Job it appears as "the Pleiades," but in Amos it appears as "the seven stars." The Revised Standard Version, English Standard Version, New American Standard Bible, and New International Version have "Pleiades" in all three verses. The Septuagint agrees with the use

---

[16] This issue is highly debated, as the orthography of Job suggests it is from Israel's monarchial period; however, the various theological indicators in the book suggest that the events it describes are much more ancient.

in the two verses from Job, but apparently lumps the Pleiades and Orion into "all things" in Amos 5:8. The Hebrew word translated "Pleiades" in all three instances is כִּימָה (*kîmāh*), meaning "heap" or "accumulation." This name is appropriate, for the Pleiades appear to the naked eye as a little lump of stars. Is there any great soteriological meaning in this? It would not appear so, given its apt description of the Pleiades. However, advocates of the gospel in the stars generally ignore the biblical (Hebrew) name for the Pleiades, opting instead for the pagan name of possibly later origin. They claim that the name "Pleiades" means "the congregation of the judge or ruler." Bullinger goes on to say this "comes to us through the Greek Septuagint as the translation of the Hebrew kimah,"[17] which is entirely without merit.

Some Responses to This Criticism

When confronted with criticism of their theory, supporters of the gospel in the stars generally have several possible responses. One approach is to appeal to the Matthew 2 account of the magi. We do not know much about the magi, but they likely did have considerable knowledge of astronomy. In most ancient Middle Eastern cultures, astronomy and astrology were intimately entwined, so it is not possible to ascertain how much the magi were involved with astrology. Supporters of the gospel in the stars ask how else the magi could have known about the birth of the Messiah unless there was a gospel in the stars of which the magi, being astronomers, must have been aware. This is a classic example of the logical fallacy of begging the question. If, as most scholars think, the magi were Persian, they likely had read Daniel's prophecy of seventy weeks and hence knew that the time of the Messiah's arrival was nigh. In fact, the magi were not the only people who were expecting the Messiah at that time, for many Hebrews were looking for the Messiah as well (see Chapter 7 on the Christmas star). We do not know what the star that the magi saw was, but supporters of the gospel in the stars want us to assume that it must have had something to do with the gospel in the stars. This is all a bit muddled though, because many supporters of the gospel in the stars believe some astronomical event, such as an unusual planetary conjunction, was the star. If that sort of thing was the star, then did the magi really require the gospel in the stars? There are, however, good reasons to believe that the star of Bethlehem was a unique supernatural phenomenon that God produced to lead the magi.

---

[17] E. W. Bullinger, *The Witness of the Stars*, 121.

Another possible response is to appeal to all the alleged parallels to gospel-related concepts found in constellations, and especially to those in the zodiac. Exhibit A is Virgo, which many take as an obvious analog to Mary. But is it? The prophecy of the Virgin Birth is from Isaiah 7:14, written a little more than seven centuries before its fulfillment. Isaiah was written long after the gospel in the stars supposedly originated, but there is no biblical evidence that there was any other prophecy or expectation of a virgin-born Messiah prior to Isaiah. Furthermore, virginity was something that was prized and much discussed in ancient cultures, as expressed by Greek and Roman poets. Given the high importance of virginity, it is not surprising that a virgin might show up among the constellations. A virgin with a child in the ancient constellations would have been quite unusual, and so perhaps might have been a good argument for the gospel in the stars. Rolleston understood this, and so she was very creative in claiming that just such a thing did exist. But, alas, she did this with quite a bit of sleight of hand, by moving an unclear figure from the Dendera planisphere to a new location and suggesting a false history to accompany it.

Or consider the presence of a ram and a bull among the zodiacal constellations. Supporters of the gospel in the stars suggest that, being sacrificial animals, these constellations must be related to the gospel story. But are they? To most of us today, these animals are a bit exotic, but they were not the least bit exotic to the largely agrarian cultures in the ancient Near East. Most people then saw these creatures nearly every day. There is nothing within these constellations to suggest that they were being sacrificed; but even if they were, what would that prove? Those animals were common pagan sacrifices. One could argue that such pagan sacrifices were perversions of God's plan, but that does not prove that the constellations are perversions of God's plan.

Other supposed parallels to the gospel story include a constellation representing a man with his foot poised over a serpent, a supposed allusion to the truth in Genesis 3:15. Being the first messianic prophecy dating from the earliest time, this approach could have merit. However, Ophiuchus has his foot over a scorpion, not a serpent. Only by the most creative touch could the scorpion be transformed into a serpent. Furthermore, there is no connection made between Ophiuchus and Scorpius in mythology. There is, however, a connection between Orion and Scorpius, although they are on opposite sides of the sky from one another. Orion has his foot above a hare. It sounds like some sort of weird magic trick to turn a rabbit into a snake, but Rolleston found a way to do that too. With the kind of loose rules of interpretation required to get that, the constellations can be turned into about anything imaginable, which, of course, proves nothing.

Mentioning Orion, when gospel in the stars advocates learn of the criticism that the biblical name for the constellation Orion means "fool," they often have a most interesting response. They argue that the name of Orion was perverted along with all the other constellation names and lore, and so the verses in Job and Amos use a perverted name for Orion. However, there are at least two problems with this response. First, Job likely is the oldest book of the Bible—it might date to as early as 2000 BC. This predates the first advent of Christ by two millennia. Yet this response requires that the magi somehow managed to preserve the gospel truth in the stars, although God's inspired Word gives no hint of it. Second, $k^e s\hat{\imath}l$ as the Hebrew name for Orion is unique to the Old Testament. Gospel in the stars advocates look for Hebrew and other Semitic language meanings in the pagan name Orion, but they totally ignore a genuine Hebrew name, because it contradicts their thesis. If God has a name for Orion, it stands to reason that He would use it in his inspired word. Ultimately, this claim about the Bible using a perverted name for Orion amounts to a patch to fix the problem, and it effectively makes the claim unfalsifiable.

Conclusions

We have found numerous problems with the gospel in the stars theory. One problem is its false history, as with the Southern Cross. Related to this is the conjecture of a supposedly lost ancient constellation, the Desired, out of the misplacement and misunderstanding of the constellation Coma Berenices. Another problem is the complete fabrication of the system of decans relating the constellations. Yet another problem is the very questionable derivation of the meanings of stars' names. This is compounded by the fact that in some instances supporters of the gospel in the stars intentionally avoid biblical names for stars, because those names would contradict the thesis. However, this approach contradicts the stated assumptions necessary for the gospel in the stars.

The concept of the gospel in the stars actually undercuts itself. The conjecture is that before the Bible was given, the patriarchs needed some way to pass along spiritual knowledge and that the constellations were the vehicle to do this. It should be obvious that this method of revelation and preservation, being subject to misunderstanding (much garbling of the original message is acknowledged by its proponents), is vastly inferior to the revealed word. If this line of reasoning is correct, then when the Bible was revealed to man, the stellar gospel ceased to serve a purpose—it had been supplanted by something better. In this late age, why would we want to return to such an imperfect vehicle? Does this alleged gospel in the stars have any significance for us today?

From Extrabiblical to Unbiblical

But there is a final, very serious objection to the gospel in the stars: it goes beyond being extrabiblical into being unbiblical. Ephesians 6:19 refers to the gospel as a "mystery." In the New Testament, a mystery is something that was previously unknown, but has now been revealed to us. Romans 16:25–26 states that this mystery was hidden for long ages and was revealed through prophetic writings (that is, in the Old Testament, not in the stars). First Corinthians 2:7–8 goes on to tell us that if the princes of this world would have known of this mystery, "they would not have crucified the Lord of glory." First Peter 1:10–12 says that while the prophets "searched diligently," they failed to grasp fully the gospel before its time. Genesis teaches us that the patriarchs knew that God required a blood sacrifice, but apparently had no idea what God's full plan of redemption would be. If they had known the full plan, proposed by the gospel in the stars, then this knowledge would have been available to the princes of this world, and there would have been no crucifixion of Jesus, So the gospel in the stars is in direct contradiction to the clear teaching of 1 Corinthians 2:7–8.

In his second epistle, Peter cautioned against cleverly devised fables (2 Pet. 1:16–21). His message here relied upon two solid foundations: (1) the eyewitness account of his legitimate experience at the transfiguration, and (2), more importantly, the testimony of Scripture

The gospel in the stars does not seem to fit into either solid foundation. With no clear teaching about the gospel in the stars in Scripture (or for that matter, prior to 1865), this sort of thing must fall under the category of fables and endless genealogies that the Apostle Paul warned against in 1 Timothy 1:4.

Christian Gnosticism

The early church had major battles with Gnosticism, and some of the New Testament epistles battled Gnostic teachings that had crept into the church in the first century. One element of Gnosticism is an emphasis on secret knowledge. That is, knowledge not generally known to the uninitiated that leads either to salvation or to some higher plane of spiritual existence. The appeal of secret knowledge is very strong, and that allure is evidenced today. Examples include the Bible code, the da Vinci code, pyramidology, ancient astronauts, and various grand conspiracy theories.

The gospel in the stars fits into this category of secret knowledge. People become aware of the gospel in the stars by reading a book or an article, hearing a sermon, or watching a video or a presentation on the topic. This information is entirely new to them; the information is not obvious; and the entire package is wrapped in references to various Bible passages. Upon learning this new information, many people feel uplifted and encouraged, though it is not clear what the reason for this good feeling is. Somehow, just acquiring this knowledge makes many Christians feel some new validation of their faith. While well intended, this new knowledge is based upon false information and is contrary to biblical principles.

Can We Salvage Anything from This?

Is there anything that we can salvage from all of this? Despite the damage wrought by purveyors of the gospel in the stars, the surprising answer is yes, we can. In his parables, Jesus used everyday examples to teach spiritual truth. The Apostle Paul did a similar thing in his sermon at Mars Hill (Acts 17:23). Paul took the inscription at a pagan shrine and launched from it a gospel message. He even quoted from Aratus, the ancient Greek poet who wrote about some of the constellations. In a similar manner, we can take constellations and draw parallels to the gospel. For instance, a discussion of Virgo can easily lead to discussion of the conception and birth of Jesus Christ.

Despite the problems with the gospel in the stars, there are two constellations that could be vestiges of primordial information regarding the messianic prophecy found in Genesis 3:15. The constellation Hercules is among the 48 original constellations, though Ptolemy did not call it that. Hercules is the Greek name for the constellation, but this name became associated with the constellation a few centuries after Ptolemy. In the *Almagest*, the constellation is a man kneeling, and all depictions show a man kneeling. Why he is kneeling is not clear, and depictions vary in certain details. These details include whether he is clothed or not, and if clothed, with what clothing. A common depiction is with lion skin clothing. In one hand he usually holds a club, but the purpose of that is not clear. In the other hand he holds various things, but frequently it is a branch with heads popping out.

In all the depictions of Hercules, one foot lies above the head of Draco. In modern times, we think of Draco as a dragon, but in ancient depictions, Draco is a snake. Indeed, in many ancient languages there is no distinction between snakes and dragons. Of course, the parallel to the first messianic prophecy found in Genesis 3:15 is striking. Furthermore, this parallel is not forced like so many other connections made by proponents of the gospel in the stars. Since this knowledge of the bruising of the Messiah's heel and the crushing of the serpent's head was known from the earliest times, it is possible that this one constellation may be a memorial to the coming Messiah, albeit with embellishments of unknown origins. It is remarkable that gospel in the stars advocates do not emphasize the possible link between these two constellations more than they do.

The End of the Matter

The gospel in the stars is such a positive-sounding teaching that it has much appeal. However, there are several arguments against its acceptance. First, it is based entirely upon conjecture and presented as a feasibility study, but with no real evidence. Second, it contains numerous factual errors which raise serious credibility problems. Third, it is extrabiblical in that it presents a doctrine that is nowhere taught in Scripture, though there are many passages in the Bible that provide ample opportunity for it to be mentioned. Fourth, by its very premise, the gospel in the stars has no relevance for those who possess the Bible, God's completed verbal revelation.

The appeal that the gospel in the stars thesis has for so many Christians is understandable. It also is understandable that many Christians find encouragement in it as well. However, we must all apply the Berean test (Acts 17:11) to the gospel in the stars theory to see whether it conforms to Scripture.

CONCLUSION

Astronomy and the Heart of the Bible

Up to this point, I have omitted one significant biblical passage that relates to the astronomical bodies—2 Peter 3:10. It reads,

> But the day of the Lord will come like a thief, and then the heavens will pass away with a roar, and the heavenly bodies will be burned up and dissolved, and the earth and the works that are done on it will be exposed.

Instead of "heavenly bodies," the King James Version says that the "elements" will burn up, but the intention is the same, for it refers to the things that are in the heavens. The "roar" here refers to the loud sound of an all-consuming fire.

T. S. Eliot's poem *The Hollow Men* concludes with these two lines:

This is the way the world ends
Not with a bang but with a whimper.

The dominant cosmology today, the big bang, posits that the universe began with a bang, but that it likely will end with a whimper, in the form of what cosmologists call a heat death. This is in stark contrast with God's word, which proclaims that the universe will end with a "bang," a sudden catastrophic conflagration.

But this is not the first time that the world will have been destroyed. In the third chapter of his second epistle, the Apostle Peter intimately tied the coming judgment of the world to the first judgment of the world during Noah's Flood. Verses 6–7 read,

> And that by means of these the world that then existed was deluged with water and perished. But by the same word the heavens and earth that now exist are stored up for fire, being kept until the day of judgment and destruction of the ungodly.

The cause of God's judgment is man's sin. While that earlier destruction of the world appears to have been confined to the earth with the purpose of cleansing the earth of man's wickedness, the coming destruction will be cosmic and more complete in that it will remove the taint of man's sin from all of creation (Rom. 8:18–23). Just as our bodies require redemption (Rom. 8:23), so the universe itself requires redemption. This will prepare the way for "new heavens and a new earth in which righteousness dwells" (2 Pet. 3:13), according to God's promise (cf. Ps. 102:25–27; Isa. 65:17; 66:22). This is similarly attested to by the Apostle John in Revelation 21:1:

> Then I saw a new heaven and a new earth, for the first heaven and the first earth had passed away, and the sea was no more.

In the third chapter of his second epistle, Peter also ties the earlier destruction of the world to the creation. Most of the third chapter is in response to a challenge from skeptics (yes, they existed even in the first century), for 2 Peter 3:3–4 reads,

> Knowing this first of all, that scoffers will come in the last days with scoffing, following their own sinful desires. They will say, "Where is the promise of his coming? For ever since the fathers fell asleep, all things are continuing as they were from the beginning of creation."

This amounts to a challenge to God's faithfulness to follow through with His promises, and ultimately is a denial of His power and even His existence. It sounds like today, doesn't it? The apostles thought that they were living in the last days, and indeed they were, albeit the beginning of the last days. How much more relevant is this challenge near the end of the last days, as we must be, since it is nearly two millennia later? Peter immediately went to the heart of the matter in his response to this challenge, as 2 Peter 3:5 says,

> For they deliberately overlook this fact, that the heavens existed long ago, and the earth was formed out of water and through water by the word of God.

As this book notes in Chapter 2, this verse (among others) underscores the vital importance of water in the creation event. Note that the source of the scoffers' problem is that they deny creation, or at the very least the role of the one true God in creation. If there is no Creator, then there is no God to judge sin, so there was no worldwide Flood. Nor will there be future judgment of the world. But the word of God is sure: there was a creation, and the waters of creation were stored up to destroy that world in the Flood. Furthermore, there is coming destruction of the world, though this time not by water, lest God should violate the covenant that He established with Noah (Gen. 9:8–17). This time "the heavens and earth that now exist are stored up for fire" (2 Pet. 3:7).

What should be our response to this message? The Apostle Peter raised this very question, for 2 Peter 3:11–12 asks,

> Since all these things are thus to be dissolved, what sort of people ought you to be in lives of holiness and godliness, waiting for and hastening the coming of the day of God, because of which the heavens will be set on fire and dissolved, and the heavenly bodies will melt as they burn!

Peter answered this question by admonishing us (2 Pet. 3:14),

> Therefore, beloved, since you are waiting for these, be diligent to be found by him without spot or blemish, and at peace.

That is, we are to look forward to this coming judgment, knowing that we will be delivered. We are to conduct ourselves in a manner that is consistent with our calling. This encouragement is given to those who are true followers of Jesus Christ, those who have placed their trust in Him for salvation.

For those who do not know the Lord Jesus Christ in this way, the message is not so encouraging. If fact, it is far from encouraging. When this judgment comes, it will be too late for those who do not know Him. It is not a matter of being a good person or even being a particularly religious person. It doesn't even matter what you believe about creation. You could agree with everything written in this book, but if you have not put your faith in Christ and in Christ alone, then it is not sufficient. In Luke 13:24–27, Jesus said,

> "Strive to enter through the narrow door. For many, I tell you, will seek to enter and will not be able. When once the master of the house has risen and shut the door, and you begin to stand outside and to knock at the door, saying, 'Lord, open to us,' then he will answer you, 'I do not know where you come from.' Then you will begin to say, 'We ate and drank in your presence, and you taught in our streets.' But he will say, 'I tell you, I do not know where you come from. Depart from me, all you workers of evil!'"

Obviously, many who believe that they are right with God in fact are not. To be right with God requires our full attention and devotion. In Matthew 13:45–46, Jesus said,

> "Again, the kingdom of heaven is like a merchant in search of fine pearls, who, on finding one pearl of great value, went and sold all that he had and bought it."

This means that salvation is so precious, that we must be willing to give up all to obtain it. However, this does not mean that we can do anything on our part to earn it or purchase it. Ephesians 2:8–9 says,

> For by grace you have been saved through faith. And this is not your own doing; it is the gift of God, not a result of works, so that no one may boast.

Sin entered into the world through the disobedience of Adam in the Garden of Eden, and sin brings death, both physical and spiritual (1 Cor. 15:21–22). But God has provided a way of salvation—a way of life—as Romans 5:8 says,

> But God shows his love for us in that while we were still sinners, Christ died for us.

Perhaps the most succinct description of God's plan of salvation is found in 1 Corinthians 15:3–4:

> For I delivered to you as of first importance what I also received: that Christ died for our sins in accordance with the Scriptures, that he was buried, that he was raised on the third day in accordance with the Scriptures.

This is the gospel in which we have our hope. All that is required is our trust in the finished work of Christ on the Cross and His Resurrection to pay the penalty for our sins and give us eternal life. If you have not done this, will you not do this today?

It is my desire that this book will encourage, edify, and educate believers. If it stirs greater love and devotion for our Creator and Savior, then it is a success. For those who do not know Him, I hope that it leads them to salvation through the gospel of Christ.

APPENDIX

Scripture as the Controlling Factor in Christian Worldview Development

Lee Anderson Jr.

Introduction: The Crucial Place of Scripture in Christian Worldview Development

A "worldview" has been defined as "a set of beliefs about the most important issues in life," or, more precisely, as "a conceptual scheme" into which an individual consciously or unconsciously places or fits everything he or she believes, and by which that person interprets reality.[1] Derived from the German word *Weltanschauung*, a worldview functions as the framework or interpretive grid through which a person attempts to make sense of every aspect of information encountered in life, and by which determinations are made about meaning and values.[2] A worldview that has been well developed is, at minimum, reflective of (or sometimes determinative of) what a person believes about God and mankind, as well as the nature of reality, knowledge, and morality.[3] It is something which every person possesses, whether or not that person recognizes it.

In view of the preceding definition, it is appropriate that systems of religion are regarded as worldviews; virtually all religions presume to speak to the subjects of God, mankind, reality, knowledge, and morality. Christianity is no different. In fact, as David Naugle rightly maintains, "Conceiving of Christianity as a worldview has been one of the most significant developments

---

[1] This useful definition is taken from Ronald H. Nash, *Worldviews in Conflict: Choosing Christianity in a World of Ideas* (Grand Rapids, MI: Zondervan Publishing House, 1992), 16.

[2] For a summary on this point, see Norman L. Geisler, *Baker Encyclopedia of Christian Apologetics*, Baker Reference Library (Grand Rapids, MI: Baker Academic, 1999), 785–86.

[3] Ronald H. Nash, *Worldviews in Conflict: Choosing Christianity in a World of Ideas*, 26–33. A worldview may concern itself with more than these things, but these are basic.

in the recent history of the [evangelical] church."[4] What, though, does it mean to have a "Christian" worldview? Ultimately, a genuinely Christian worldview entails much more than a commitment to "Judeo-Christian ethics" or even an admiration for the person and teachings of Jesus Christ. It must involve a commitment to the authority of Scripture and, indeed, be molded around the very teachings of Scripture. John MacArthur rightly states, "A truly *Christian* worldview, simply put, is one in which the Word of God, rightly understood, is firmly established as both the foundation and the final authority for everything we hold true," and, accordingly, "When we begin with a right view of Scripture, the Bible itself ought to shape what we believe from start to finish."[5]

How, though, does Scripture function to supply the substance of or define the parameters of a genuine Christian worldview? This question can only be properly answered in the light of an understanding of the theological process that feeds into worldview development. The theological task begins with *biblical theology*, that is, the analysis of the doctrinal content of each individual book of the Bible (or group of books by a single author) with due consideration given to its place in the history of God's progressive revelation. This is followed by the integration or synthesis of the messages of individual texts across the boundaries of history and authorship. This leads to an organization of biblical truth in a categorical or *systematic* fashion,[6] which, in turn, provides the doctrinal basis for the validation and incorporation or (alternatively) the rejection of extrabiblical truth claims that arise out of the study of history, science, and other disciplines. The final stage in the theological task involves the application of the resulting worldview to all aspects of life. The goal of this process is that the Christian worldview be biblically consistent throughout.

---

[4] From David K. Naugle, *Worldview: The History of a Concept* (Grand Rapids, MI: William B. Eerdmans Publishing Company, 2002), 4. Naugle notes that although the word "worldview" is of recent origin, "such a grand, systematic vision of the faith is not." He goes on to assert, "It has a distinguished genealogy, going all the way back, of course, to the Bible itself with its doctrine of a trinitarian God who is the creator and redeemer of heaven and earth and whose sovereignty rules over all" (p. 5). Christianity as a worldview was further developed by the luminaries of the early church, the medieval theologians, the Protestant Reformers, the English and American Puritans, and, later, the evangelical community—especially in North America.

[5] MacArthur likewise refers to Scripture as "the standard by which we must test all other truth-claims." This axiom is fundamental to right Christian thinking. See John MacArthur, "Embracing the Authority and Sufficiency of Scripture," in *Think Biblically: Recovering a Christian Worldview*, edited by John MacArthur, Richard L. Mayhue, and John A. Hughes (Wheaton, IL: Crossway Books, 2003), 21.

[6] Notably, the main categories (e.g., bibliology, Christology, etc.) for the systematic organization of biblical truth are not extrabiblical, but emerge out of the discipline of biblical theology. These doctrinal categories represent dominant themes which run throughout the text of Scripture.

Indeed, the single most critical factor in the development of a Christian worldview is that it must be wholly biblical. This demands not just that Scripture contributes to the development of the worldview, or that it supplies the basic content of the worldview (however true this must be), but that Scripture contributes directly to each progressive stage of the theological task and that it ultimately serves as the primary *control* in the construction of the worldview from the beginning to the end. In support of this thesis, this essay will consider the manner in which the word of God must be approached in order to develop a robust Christian worldview. It will then briefly survey the various stages of the theological task so as to show that the Scriptures directly intersect with each step, and that they contribute to the continuing refinement of the Christian worldview.

The Prerequisite to Approaching Scripture in Christian Worldview Development

In acknowledging that Scripture is foundational to the development of the Christian worldview, there is an inherent requirement that one approach the Scriptures in the correct manner in order to arrive at a worldview that is genuinely consistent with what Scripture teaches. Proverbs states that the "fear of the LORD" (יִרְאַת יְהוָה) is the "beginning" (or "first principle") of both knowledge and wisdom (Prov. 1:7; 9:10; cf. 15:33; Job 28:28; Ps. 111:10). The fear of the LORD is therefore essential to cultivating a biblical mindset and—by derivation—to developing a sound and robust Christian worldview.[7] This is where a correct perspective on reality begins. D. Bruce Lockerbie rightly surmises, "Wisdom and knowledge, not reason and intuition, are the goal of all cognition, all learning, all thinking. And the beginning point is an obligatory reverential awe before God the Father Almighty, Maker of heaven and earth."[8]

---

[7] Note on this the observations offered by Richard L. Mayhue, "Cultivating a Biblical Mind-set," in *Think Biblically: Recovering a Christian Worldview*, edited by John MacArthur, Richard L. Mayhue, and John A. Hughes (Wheaton, IL: Crossway Books, 2003), 53. Naugle likewise argues, "From a scriptural point of view…the heart is responsible for how a man or woman sees the world." He continues on to say, "The Bible itself asserts that the inner life must be rightly aligned with God and have the appropriate attitude of reverence if it is to receive insight into the divine wisdom which orders the cosmos." David K. Naugle, *Worldview: The History of a Concept*, 272. See also on the biblical presentation of wisdom in comprehending reality Garrett J. DeWeese, *Doing Philosophy as a Christian*, Christian Worldview Integration Series (Downers Grove, IL: IVP Academic, 2011), 52.

[8] D. Bruce Lockerbie, "Thinking Like a Christian, Part 1: The Starting Point," *Bibliotheca Sacra* 143, no. 569 (1986): 9. In a similar fashion, Brad Green contends that the biblical concept of the "fear of the LORD" is basic to informing a Christian perspective of how faith properly intersects with understanding (or reason). The notion of "faith seeking understanding" is, as Green argues, fundamental to a Christian education, which is, or at least ought to be, aimed toward constructing a Christian worldview. Brad Green, "Theological and Philosophical Foundations," in *Shaping a Christian Worldview: The Foundations of Christian Higher Education*, edited by David S. Dockery and Gregory Alan Thornbury (Nashville, TN: B&H Academic, 2002), 80–81.

In Scripture, the concept of the fear of the LORD is closely associated with hearing and receiving the word of God (Deut. 4:10; 17:19; 31:12–13). This "fear" involves an awe-filled reverence toward God in view of a knowledge of who God is and what He has done (Deut. 28:58; cf. Pss. 22:23; 33:6–9; 47:2; 64:9) which, in turn, produces a dependence upon God's revelation in navigating the path of life (Prov. 2:1–5; 3:5–7). Indeed, in light of verses that showcase man's finiteness and limited wisdom by comparison to his Creator (e.g., Job 28:13–28; 37:21–24), it must be concluded that the fear of God entails a recognition of the fact that God surpasses the limitations of human wisdom, followed by a response on the part of man to God's demand that man listen to Him.[9] This observation is vital to what follows.

The essential attitude of reverence which is inherent in the fear of the LORD dictates that the theologian must approach Scripture so as to receive from Scripture what it intends to teach in developing a worldview, rather than attempting to fit Scripture into a preconceived worldview or (worse) imposing on Scripture an alien worldview. However, as Daniel Castelo underlines, "The 'fear of the Lord' is not simply a principle affirmed at the beginning of a theological treatise that informs the way an emerging system develops. Rather, the fear of the Lord is the disposition that sustains and maintains the task of theological reflection as legitimately theological."[10] As such, it is the fear of the LORD that compels the theologian to interact with the text of Scripture at every stage of the theological task as he moves toward the crystallization of a Christian worldview. So too, it is the fear of the LORD that compels submission to the Scriptures in the application of that Christian worldview to every dimension of life. Indeed, the fear of the LORD is an *indispensable component* in the process of shaping and implementing a Christian worldview.

The Continuing Employment of Scripture in Christian Worldview Development

A defining mark of evangelical Christianity is that "the Bible alone, and the Bible in its entirety, is the Word of God written and is therefore inerrant in the autographs."[11] Because the Scriptures are, in their entirety, given by inspiration of

---

[9] The author wishes to thank Abner Chou, professor of Biblical Studies at The Master's College, for his insights which helped lead to this explanation of the "fear of the LORD."

[10] Daniel Castelo, "The Fear of the Lord as Theological Method," *Journal of Theological Interpretation* 2, no. 1 (2008): 158. Castelo's view on theological method does not necessarily conform to the particulars of the steps mentioned in this essay, but his point is still fundamentally correct.

[11] Notably, this statement, as well as confessed agreement with the orthodox doctrine of the Trinity, is the Doctrinal Basis for the Evangelical Theological Society (*ETS Constitution*, Article III).

God or "God-breathed" (2 Tim. 3:16), and since they are therefore infallibly true (cf. Ps. 119:160; John 17:17), it may be boldly affirmed that they are authoritative—not just in matters of "faith and practice," but in every area to which they speak. It is on this basis, therefore, that the Scriptures may be claimed as the "appropriate beginning point for shaping a worldview."[12] The notion of looking to the Scriptures as a "starting point" does not, however, suggest that they are merely a source to be examined toward the beginning of Christian worldview development, but can subsequently be put aside.[13] As it concerns the task of theology, the Scriptures do not reside, as it were, upon a horizontal axis that can be left behind; rather, they are situated upon a vertical axis on which all else rests. Consequently, the Scriptures are not just the logical beginning point for worldview formulation; they also inform the development of the Christian worldview at every subsequent step along the way. This continued interaction with the text of Scripture will, ideally, spare the theologian from straying into doctrinal error.

Generally speaking, the most obvious dependence that the Christian theologian will have on the Scriptures evidences itself in the first stage of theological development, that is, at the level of biblical theology. Notably, this stage in the process of theology and worldview development is preparatory for all that follows; it is where the theologian makes a concerted effort to determine what each individual text of Scripture is communicating in light of its author's historical context and in a manner consistent with the lexical, grammatical, and structural elements of the passage. This step of the theological process can only be rightly executed if there is a commitment to read Scripture in a *normal* sense, that is, with a consistent literal, grammatical-historical hermeneutic, making room for symbolism and other obvious figures of speech in the text.[14] This step is clearly foundational; it is what supplies the basic *substance* of a Christian worldview. Obviously, if the Scriptures are ignored at this point, the only alternative is to look elsewhere for the basic content of one's worldview, which defeats the idea of the worldview being distinctly *Christian*.

---

[12] This basic point was expressed by George H. Guthrie in his essay entitled, "The Authority of Scripture," in *Shaping a Christian Worldview: The Foundations of Christian Higher Education*, edited by David S. Dockery and Gregory Alan Thornbury (Nashville, TN: B&H Academic, 2002), 28. Essential to Guthrie's explanation of this point is his exposé of inroads made by postmodernism into Christian thinking (pp. 28–30).

[13] To be clear, Guthrie's essay does not imply that such a perspective is valid.

[14] See on this especially the points by Ron M. Johnson, "Systematic Theology and Hermeneutics," in *An Introduction to Classical Evangelical Hermeneutics: A Guide to the History and Practice of Biblical Interpretation*, edited by Mal Couch (Grand Rapids, MI: Kregel Publications, 2000), 32–35. Johnson's pointed assertion that one's hermeneutics determines his or her theology must not be taken lightly.

Beyond this, however, the Bible needs to continue to have an integral role in theological formulation and worldview development. In integrating the messages of individual texts across the boundaries of history and authorship, for example, the synthetic conclusions that are reached by the theologian must remain subject to the plain meaning of each individual text. Accordingly, the work of arriving at the integrated meaning must not be allowed to distort the sense of a given text read within its historical and literary context. The theologian is not free to offer conclusions about the integration of two or more passages that undermine the meaning present in any one of them. Naturally, this, like the preceding stage of the theological process, requires a regard for the principle of literal interpretation; texts must not be artificially allegorized or "spiritualized" just to make a proposed synthesis "fit." Also vital at this stage is a healthy regard for the progress of revelation. The sense of a later text cannot be imposed upon an earlier passage so as to override or change the meaning of the earlier text. Consequently, it may be concluded that *the Bible itself* provides the control for the synthesizing of all biblical content in the process of theology. Also, in moving from the synthesizing of biblical content onto the categorizing of biblical content, the Bible itself, once again, functions in a controlling capacity, in that it is the principal or dominant themes of Scripture that serve as the main doctrinal categories for systematic theology.

Subsequent to categorization, it is in the next stage of the theological process (that is, the validation or invalidation of extrabiblical truth claims) that attention to what the Scriptures teach becomes pivotal. Significantly, even the Bible itself directs its readers to extrabiblical sources to corroborate its truth claims. The biblical text affirms, for instance, the value of natural revelation (Rom. 1:18–21) and the witness of history (1 Cor. 15:6). Nevertheless, the inherent authority of Scripture means that, in the task of worldview development, Scripture must sit in judgment over claims from the study of science, history, and other disciplines—and not vice versa.[15] The Bible, and the Bible alone, is "perfect" (cf. Ps. 19:7–9; James 1:25), and so qualified to be the standard against which other truth claims are evaluated, warning believers away from untrustworthy and empty philosophies (cf. Ps. 19:10–11).[16] Accordingly, at this stage of the theological process, the Scriptures serve to protect the emerging Christian worldview, functioning to filter out the claims which are inconsistent with what the Lord God Himself declares to be true and right.

---

[15] This point is largely a restatement of observations made by John M. Frame in his treatment *Apologetics to the Glory of God: An Introduction* (Phillipsburg, NJ: P&R Publishing, 1994), 60, cf. pp. 57–59.

[16] John MacArthur, "Embracing the Authority and Sufficiency of Scripture," 29–33.

Continuing on to the final stage of the theological process, the application of the Christian worldview to every dimension of life, the Scriptures must remain in a position of control. At this juncture, the main question that arises is "How *should* the Christian *think*, *speak*, and *act*?" This question's answer cannot be arrived at merely on the basis of personal preference or sentiment, or on the basis of cultural or political acceptability. Ultimately, everything that has been gained from all of the preceding stages of theological formulation and worldview development must be brought to bear on answering this question, lest there be a dreadful disconnect in the *purpose* of the theological task. It deserves to be noted that the proper application of a Christian worldview involves sensitivity concerning, among other things, the marked differences between descriptive and prescriptive texts, the similarity or dissimilarity in the circumstances of the original readers and contemporary readers, and the progress of revelation—especially as it relates to distinctions between Israel and the Church. However, the fact remains that Scripture is key to every stage of theological formulation, and that the implementation of that theology suffers tremendously if the Scriptures are not likewise central in the final step of application to the Christian life.

The Unending Refining Use of Scripture in Christian Worldview Development

This essay has demonstrated, in brief, that the Scriptures are essential (and indeed, foundational) to every step in the process of theological formulation and Christian worldview development. To assume, however, that this process is strictly linear is not accurate. Rather, the development of a Christian worldview is (or at least should be) characterized by a complex web of feedback loops that lead to continuing refinement in that worldview.[17] This is to be expected for two reasons: (1) the inexhaustibility of the contents of Scripture, and (2) the continually changing world, with its relentless introduction of new truth claims—all of which must be subjected to the scrutiny of the text and either affirmed for incorporation into the Christian worldview or rejected.

---

[17] The fact that the Christian worldview ought to be subject to continued refinement means, naturally, that while it is possible to have a truly robust Christian worldview that is rigorously biblical in all its aspects, never will believers (at least on this side of glory) arrive at a *totally* comprehensive worldview that is *perfect* in all respects; a believer's worldview will always need further honing in the light of Scripture.

The most obvious means by which refinements will be made to the Christian worldview comes back at the level of biblical theology, the first stage of theological formulation. Indeed, as the Christian continues to study Scripture, giving attention to the lexical, grammatical, structural, and contextual aspects of the text, he or she will (by the aid of the Holy Spirit) gain an increased understanding of what the Scriptures mean. This will inevitably lead to adjustment in a person's biblical theology, which may carry with it ramifications (whether discrete or substantial) for the later stages of the theological process. Additionally, challenges arising at the levels of synthesis or integration, categorization or systematization, and the validation/ invalidation of extrabiblical truth claims may bring to light inconsistencies and flaws that demand a reworking of theology at the lower levels. Even at the level of the application of the Christian worldview to all aspects of life, the possibility remains that weaknesses will evidence themselves, requiring further study of the text, with the desired outcome of making one's worldview more comprehensive in scope, as well as more consistently biblical—more distinctly *Christian*—in all of its elements.[18]

A question remains concerning the extent to which extrabiblical truth claims from fields such as history and science may be allowed to influence change in the formulation of a Christian worldview. Space does not permit a full investigation of this question. However, it deserves to be said that while human interpretations coming from history, science, and other fields of study may never rightly be used to dismiss the clear truth claims of Scripture, they should, where they seem to clash with Scripture, spur the theologian back to the study of the text, to ensure that he or she has correctly worked through all of the many aspects of exegesis, synthesis, and systematization. This revisiting of the theological process will help to strengthen the Christian worldview. In this way again, the Scriptures are shown to be the controlling factor in the development of a Christian worldview. The Scriptures must serve to filter the influences of extrabiblical truth claims.

The main point is that the Christian worldview is not merely a set of facts to be accepted, but an interconnected web of truth to be explored, delineated, expounded, and applied. There is always a need for fine-tuning in the Christian's worldview as the Christian advances in the study of God's word and

---

[18] It should be noted that, notwithstanding the supremacy of Scripture in Christian worldview development, as well as the role that other fields (history, science, etc.) have in helping to flesh out the details of a worldview, the question of whether or not a person can live consistently with his or her worldview cannot be ignored. Application is indeed an important part of worldview development for Christians.

continues to engage with God's world. It is to be expected that, in the course of a Christian's life, his or her worldview will undergo significant change—hopefully becoming broader and deeper along the way. Ultimately, changes and refinements should always be for the purpose of moving the worldview towards greater consistency with the truth of Scripture.

Conclusion: The Centrality of Scripture in Christian Worldview Development

This essay has contended that attention to the Bible—the word of God—is vital to the process of theological formulation and the development of a Christian worldview. In order for a worldview to be truly *Christian*, Scripture must be more than a contributing factor; it must be the *principal, controlling* factor in the construction of that worldview from the beginning to the end. Scripture must be approached with a proper attitude (in the "fear of the LORD"), which produces a genuine receptivity to its teachings and submissiveness to its authority. Scripture thus serves to supply the basic content for theology, which, when properly synthesized and systematized, functions as the standard against which all other truth claims are evaluated. Scripture is, similarly, the primary standard for determining the appropriate application of the emerging Christian worldview to all aspects of life. Accordingly, Scripture intersects with every stage of theological formulation. It must therefore be regarded as *central* to the development of a Christian worldview.

Subject Index

atmospheric extinction 154

Augustine 249

Auriga 295

autumnal equinox 78, 85, 90, 95, 96, 173

Aviv 79, 94–97

Azariah 103, 106

Azekah 117, 118, 120

Babylon 62, 63, 102–104, 106, 123, 124, 139, 167, 169, 178, 186, 187

Bands 280

Banks, William D. 261

Barrack 120

Bartsch, Jakob 283

Bates, Joseph 73

Battle of Hastings 135

Battle of Qarqar 34

Battle of the Philippine Sea 252

Bauval, Robert 72

Bayer, Johann 283, 285

beast, the 62, 186

Bede 264, 304

Beelzebul 180

Berenice II 285

Betelgeuse 294

Beth Alpha Synagogue 308

Beth-horon 117

Bethlehem 114, 131, 132, 137, 138, 309, 310, 313

Bible code 317

biblical theology 15, 21, 23, 324, 327, 330

big bang 19, 42, 54, 200, 319

Big Bear 68, 74, 296, 297

Big Dipper 68, 74, 297

Biltz, Mark 173, 175–177

binary star 66, 206

biogenesis 223

black hole 214

blood 33, 79, 110, 152–155, 170–174, 176–178, 227, 316

blood moon 155, 173, 174, 176, 177

bloodletting 109, 110

Book of Abraham, the 71

Book of Denkard 193

Boötes 74, 296

Boyd, Steven W. 29, 44

Bradley, James 255

Brendan of Clonfert 239

Brown, Walt 41, 197, 198

Bruce, F. F. 25

Bul 97

Bullinger, E. W. 260, 261, 285, 286, 288, 290, 293, 295, 297–299, 301, 307, 310–313

Bundahis 192, 194, 195

Burgess, Stuart 19

Buxtorf, Johannes 302

Cain 277

calendar 61, 62, 84–90, 92, 94, 95, 97, 99, 127, 143–145, 148, 176, 181, 183, 185, 188–195

Cana 205, 207

canals (on Mars) 16, 17

Cancer 257, 258, 282

Canis Major 306

Canis Minor 306

canopy 41, 52

Capella 74, 295, 296

Capernaum 205

Capricornus 257, 282, 293

Carina 264

Carmeli, Moshe 215

Cassuto, Umberto 46, 189

Scripture Index